Glimpsing an invisible universe

Glimpsing an invisible universe

THE EMERGENCE OF X-RAY ASTRONOMY

RICHARD F. HIRSH

Assistant Professor, Department of History
Virginia Polytechnic Institute and State University

The right of the
University of Cambridge
to print and sell
all manner of books
was granted by
Henry VIII in 1534.
The University has printed
and published continuously
since 1584.

CAMBRIDGE UNIVERSITY PRESS

Cambridge

London   New York   New Rochelle

Melbourne   Sydney

CAMBRIDGE UNIVERSITY PRESS
Cambridge, New York, Melbourne, Madrid, Cape Town, Singapore, São Paulo, Delhi

Cambridge University Press
The Edinburgh Building, Cambridge CB2 8RU, UK

Published in the United States of America by Cambridge University Press, New York

www.cambridge.org
Information on this title: www.cambridge.org/9780521312325

First published 1983
First paperback edition 1985
Re-issued in this digitally printed version 2009

A catalogue record for this publication is available from the British Library

Library of Congress Catalogue Card Number: 82–23633

ISBN 978-0-521-25121-1 hardback
ISBN 978-0-521-31232-5 paperback

Contents

Acknowledgements

The generous support and encouragement of many people meshed to create this book. P. Frank Winkler, my former physics professor at Middlebury College, initially spurred my interest in X-ray astronomy. A practitioner in the field, Frank repeatedly urged me to do 'something honest' in life and become a physicist. Instead, I wrote about his specialty. I conducted my first historical research on X-ray astronomy when assisting Homer E. Newell, Jr, a former National Aeronautics and Space Administration (NASA) administrator, in writing a book describing his 30 years as a director of space research. While working at NASA for two summers, I was unselfishly aided by Monte D. Wright, Director of the agency's History Office, and by Frank W. Anderson, Alex F. Roland, Carrie E. Karegeannes, Lee D. Saegesser, Nancy Brun, and Martha Parr.

A fellowship from the Smithsonian Institution's National Air and Space Museum allowed me to pursue further research. My advisers, Tom D. Crouch and Paul A. Hanle, assisted me greatly in understanding the influence of technological innovation on the growth of modern scientific specialties. But perhaps I benefited most by interacting with research fellows having different academic backgrounds. Robert Marc Friedman, for example, emphasized the value of simple and direct writing. I only hope I have mastered enough of his lessons for this book. And Joseph J. Corn contributed useful advice on organizing the manuscript, often during our daily jog past tourists and protestors on the Washington Mall. My positive experience at the museum was enhanced by the kind assistance of Diane Palmer, Dominick Pisano, Catherine Scott, Phyllis Dobson, Robert Van Der Linden, Philip Edwards, Peter Suthard, and Mimi Scharf in the library, and by the aid, patience, and friendship offered by Frederick C. Durant, Howard Wolko, Walter Flint, Louis Purnell, Richard Hallion, Frank Winter, Walter Dillon, Gregory Kennedy, Barbara Pawlowski,

Diane Pearson, Louise Hull, Wendy Ackerman Stephens, Lillian Kozloski, Holly Laffoon, and Nancy Byrkit.

For aid in using their institutional resources, I am indebted to David K. Allison of the Naval Research Laboratory, Spencer R. Weart of the American Institute of Physics, Henry Small of the Institute for Scientific Information, and Richard Hart of the National Academy of Sciences. Much of the book's discussion on X-ray astronomy's social structure benefited from data and insights provided by Thomas F. Gieryn of Indiana University, whose assistance is gratefully acknowledged. I also appreciate the time and effort expended by the scientists, engineers, and administrators who consented to interviews as part of my research (see the bibliographic note on unpublished sources).

For continuing advice and encouragement while writing my dissertation, an early version of this book, I am grateful to my adviser at the University of Wisconsin, Daniel M. Siegel. His characteristic good cheer and sound criticism helped me produce better pieces of scholarship. Further valuable comments came from Aaron J. Ihde and Terry S. Reynolds. My present colleagues at Virginia Polytechnic Institute and State University also deserve special thanks. Lisa Donis, Connie Aikens, Carolyn Alls, and Linda Arnold taught me how to use our new 'toy,' a word processor, which provided tremendous freedom and latitude in revising the manuscript. Many thanks are due to Harold C. Livesay for advising me on the business aspects of publishing, and to Daniel B. Thorp and Carla Knapp, who always know how to celebrate scholarly accomplishments with impromptu parties and a 'fresh' bottle of wine. As friend and as scholar, Albert E. Moyer carefully examined the text and highlighted logical inconsistencies that I hope have since been eliminated. And from a faraway outpost in Orlando, Florida, the manuscript received scrutiny from the journalistic (and lovely!) eyes of Lisanne Renner. Bert and Lisanne's efforts surpassed the bounds of professional courtesy and bolstered my courage to publish this work.

Finally, I owe much to my parents, sister, and grandmother for encouraging my intellectual development. I dedicate this work to these loving people.

Introduction

X-ray astronomy reveals an invisible universe. Having evolved beneath an atmosphere that absorbs incoming X-rays, human beings are blind to the cosmic phenomena that produce the emanations. But in 1962, Geiger counters launched above the atmosphere caught the first glimpse of an X-ray star. Spewing forth X-rays 100 million times faster than the Sun, the extraordinary object intrigued scientists. To explain the huge emissions, investigators discarded classical theories of stellar energy production and resorted to esoteric concepts that encompassed rotating neutron stars and black holes. Though now common in the professional and popular press, these ideas bordered on science fiction 20 years ago. As it appears today, the invisible X-ray universe is very different from the one observed through unaided eyes: it is a cosmos of explosive high temperatures, intense gravitational fields, and rapid time variations.

This book deals with the evolution of X-ray astronomy during the field's initial phase of development. The period of interest begins in the late 1950s when scientists first considered studying high-energy radiations emitted by celestial objects other than the Sun. After the discovery of an unexpectedly bright X-ray source in 1962, theorists and experimentalists quickly entered the field and strove to answer the central question: what physical processes cause celestial X-ray emissions? In 1972 scientists demarcated the field's development as their thinking coalesced around a conceptual framework for resolving the major part of this problem. Armed with some paradigmatic principles, X-ray investigators in the next stage began exploring questions of interest to the broader discipline of astronomy.

In examining only the early period of the field's history, when scientists acquired fundamental data and basic theoretical principles, the book does more than simply document the accomplishments of investigators working to answer a well-defined question. Such a discussion would be inadequate.

1

It would not explain why the field emerged when it did, nor how scientists gained support for making and interpreting their observations. To understand these historical dimensions, one must realize that the investigations were intricately related to concomitant advances in research technology and to a political program designed to bolster the declining prestige of the United States. The book, therefore, also explores how technological innovation and broad public policies played important roles in the emergence of X-ray astronomy.

Beyond describing the emergence of a specific scientific field, this history of X-ray astronomy highlights three interrelated transformations that characterize the entire astronomy discipline in the United States since World War II. First, it demonstrates that the classical task of studying the cosmos in optical wavelengths has been augmented by observations in other spectral bands. As seen in Figure 1, the atmosphere provides only a few windows in the electromagnetic spectrum for terrestrial observers. Other than optical wavelengths (between about 4000 Å and 7000 Å), ground-based astronomers can only examine portions of the radio and infrared spectra. The X-ray region, for example, lying between 0.1 Å and 100 Å, is totally obscured. Taking advantage of new electronic technologies developed during World War II, radio astronomers were the first to broaden the perspective of astronomy by observing a highly energetic universe that was invisible to classically trained optical astronomers.[1] Soon thereafter, adventurous physicists installed radiation detectors on rockets that flew above the absorbing atmosphere and viewed the Sun and stars through the ultraviolet window of the spectrum. As the space race of the late 1950s and 1960s intensified, astronomical studies from high-altitude vehicles diversified into the gamma ray, infrared, and X-ray regions, making astronomy an all-wavelength activity. Of these areas of spectral research, X-ray astronomy gained the largest following in the shortest time span.

The second major change in modern astronomy consisted of the government's preemption of private institutions as the predominant funding source. In 1972, for example, federal support for astronomy amounted to $125 million, putting it in the league of nuclear physics and other 'big' sciences. Though this amounted to about one-third of funding for physics research, government spending for astronomy nevertheless eclipsed private sector allocations of $94 million for *both* astronomy and physics.[2] Beginning soon after World War II, this support initially came from military services, which supplied grants for basic investigations and new research technologies.[3] Later, by creating the National Science Foundation in 1950 and the National Aeronautics and Space Administra-

Figure 1. Attenuation of electromagnetic radiation in the atmosphere. Solid curves indicate the altitude (and corresponding pressure as a fraction of one atmosphere) at which the indicated fractional attenuation occurs for radiation of a given wavelength. Along the top of the diagram are the conventional designations of the different wavebands. X-rays are bracketed by far ultraviolet and soft gamma radiation. From M. S. Longair, *High Energy Astrophysics* (Cambridge University Press, 1981), p. 90.

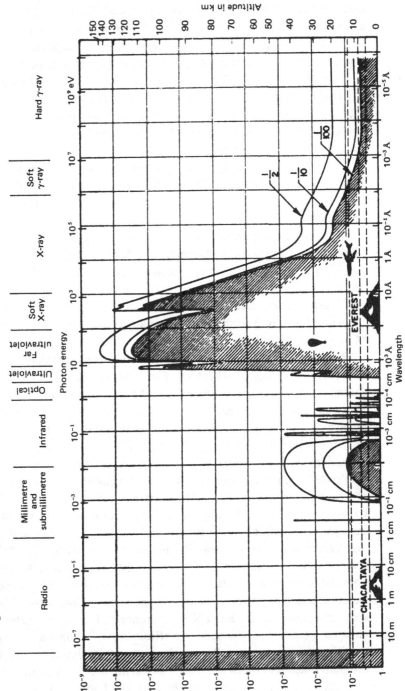

tion (NASA) in 1958, the government provided still more benefactors for astronomers. This enhanced support reflected a new public policy aimed at producing a store of fundamental scientific knowledge and technically trained people for use in national emergencies. Americans perceived such an emergency when the Russians launched the Sputnik satellite in 1957. While the government initiated no public policy specifically for X-ray astronomy, the specialty benefited in the 1960s as an unintended consequence of the massive support for space exploration following the Russians' space spectacular.

Transformations in the scope and funding of astronomical investigations contributed to the third significant change in the discipline: the greatly altered size and social structure of the research community. The old guard optical astronomers, trained to work in seclusion with existing instruments, were joined by a growing number of technologically oriented radio engineers and physicists collaborating in large teams. Only slowly did classically educated astronomers become accustomed to working with the 'migrants' in a synergistic manner. X-ray astronomers were typical of the new breed of star gazers. Trained in experimental fields of physics, they 'invaded' the new specialty and immediately constituted a large fraction of publishing scientists.

The analysis of X-ray astronomy's evolution is organized into four sections comprising 11 chapters. The first section of three chapters describes the scientific, technological, and political environments in which X-ray astronomy emerged. In retrospect, one can see that several areas of high-altitude research, such as ionospheric physics and solar X-ray astronomy, were forerunners of the new field. They provided the research technologies, in the forms of rockets and detectors, that would be adopted wholesale in X-ray astronomy. They also instituted a scientific tradition of investigations in and above the atmosphere that would be continued by later practitioners. By examining X-ray emissions from the Sun, for example, scientists obtained their initial – albeit limited – look at the high-energy universe. In view of knowledge gathered in the 1960s, the solar observations appear ironic, because they implied that X-rays from nonsolar sources would be feeble and undetectable. But the political situation created by Sputnik prompted people to explore original fields of space investigation, including X-ray astronomy. Despite predictions that augured poorly for the specialty, Riccardo Giacconi, a young and ambitious scientist at a small research company, developed an exploratory program that led to the detection of the first intense X-ray emitter in 1962.

In the second section of three chapters, the book discusses the social structure of the X-ray astronomy community. It begins with a description of how Giacconi's team encountered competition from other groups even before the first discovery. Herbert Friedman, a long-time rocket scientist bouyed by vast resources at a government research laboratory, led one competing team. A classic scientific rivalry immediately developed, spurring both groups to pursue imaginative research throughout the 1960s. Meanwhile, still more scientists entered the field because they saw the immediate and potential implications it had for resolving cosmological problems and for altering views of the universe. Migrating from physics specialties and with the support from NASA and some untraditional sources, these investigators made X-ray astronomy the fourth largest astronomy specialty by 1972.

The four chapters in the penultimate section examine attempts made during the 1960s and early 1970s to resolve the field's major problem concerning the physics of X-ray sources. They discuss the research strategies of the leading groups and how the programs and advanced technologies led to discoveries of unusual features of X-ray emitters. The chapters further outline how concurrent events in theoretical astrophysics and radio astronomy provided new avenues for research on such esoteric objects as neutron stars. Despite these efforts, the nature of most celestial X-ray phenomena remained a mystery in 1970. By the end of 1971, however, the conceptual deadlock in the field began to weaken. Giacconi and his colleagues provided the critical advance with a satellite named 'Uhuru,' which effectively ended the 'glimpsing' phase in the field's history and ushered in one characterized by long-term observations. Acquiring crucial data that could not be garnered by rocket-borne instruments, investigators resolved the central problem by suggesting a model that considered most galactic X-ray sources as binary systems consisting of a compact star and a large companion. As the small body – either a neutron star or a black hole – attracted matter from its partner, X-rays would be produced. Assimilating most observational facts gathered in the first 10 years of experimental research, this framework also served as the basis for understanding newly detected X-ray phenomena during the next phase of its evolution.

The final section, an epilogue, reviews trends in technology and public policy that affected the first phase of X-ray astronomy's evolution. It describes the importance of public policy in supporting advances in research technology, which in turn stimulated dramatic progress in conceptual understanding. It also documents the impact on the specialty of a nebulous public policy in the 1970s – a policy that no longer promised the consistently high level of funding enjoyed in the previous decade. Instead of being a favored group of investigators, X-ray astronomers constituted one of many political interest groups vying for a slice of the federal budget pie. Consequently, the scientists learned that their dependent relationship on the government acted like a double-edged sword: the governmental support incubated their field and established it as a major force in the discipline. But once built up, the relationship led to disappointments when political sentiments changed. Of course, vacillating public policies have affected several scientific specialties besides X-ray astronomy. For this reason alone, the history of the field reveals much about the way science evolves – or stumbles along – in the modern United States.

SECTION I

THE SCIENTIFIC, TECHNOLOGICAL,
AND POLITICAL ENVIRONMENTS

1

The heritage of X-ray astronomy

X-ray astronomy is a gift of technology. Without the rocket technology developed in the first half of the twentieth century, instruments could not have been lifted above the attenuating atmosphere. And without specialized detectors used previously for observing cosmic radiation, scientists would have had difficulty studying nonsolar X-rays. But beyond this technological heritage, X-ray astronomy belongs historically to a class of specialties in physics and astronomy that was concerned with the upper atmosphere. At high altitudes, where the air is thin, physical phenomena occurred that held practical significance for terrestrial observers. The upper atmosphere was also the site from which cosmic phenomena could be studied most effectively. X-ray astronomy is, therefore, one of the more recent descendants of work done by scientists – mostly physicists – who lofted instruments into the upper atmosphere and beyond.

Meteorology and cosmic ray physics

The prehistory of X-ray astronomy can be traced back over 200 years to the first high-altitude research on the weather. As early as 1749, the Scotsman Alexander Wilson rigged a kite, a 2500-year-old Chinese invention, to obtain atmospheric temperatures at different altitudes. Systematized meteorological research began in the 1880s and 1890s, when effective box kites first carried self-recording instruments. Soaring up to a height of 7000 meters, the kites flew regularly for the US Weather Bureau. By the 1930s, however, they had become a hazard to the increasing number of aircraft flying the skies, and the Bureau discontinued their use.[1]

Developed in the 1780s, hot air and hydrogen balloons proved more successful than kites for meteorological studies. Their value was demonstrated on 1 December 1783, when the French physicist J. A. C. Charles, author of the gas law bearing his name, rose in a balloon to an altitude of about 3000 meters, where he measured a temperature that was 12 degrees Celsius lower than on the ground.[2] This first ascent for scientific purposes impressed other investigators, including the chemist Antoine-Laurent Lavoisier, who wrote that it showed how researchers 'can rise up to the clouds to study the cause of meteors [i.e., the cause of the weather].'[3] By the 1870s scientists realized they did not need weighty human operators to measure the temperature, pressure, and other atmospheric features at high altitudes. Using vehicles known as 'sounding balloons,' a term derived from seamen who used sounding lines to measure the unknown depths of waters, scientists could send instruments to oxygen-poor altitudes of about 30 km. In 1898, Leon Teisserenc de Bort of France and Richard Assman of Germany exploited balloon techniques and independently discovered the stratospheric region of the atmosphere, where the temperature remained constant – instead of falling – at altitudes above 10 km.[4] This observation constituted 'the most surprising discovery in the whole of meteorology' in the words of the prominent early twentieth century meteorologist, Sir Napier Shaw.[5] Since the discovery, sounding balloon measurements of the stratosphere became routine.

In contrast to meteorologists interested in occurrences *in* the atmosphere, another group of scientists used the atmosphere as a location for studying phenomena *beyond* it. These investigators examined high-energy extra-terrestrial radiations, and their 'cosmic ray' research became a major activity within the experimental physics community. Following 1912, when the Austrian Victor F. Hess determined that nonsolar radiations struck the Earth from beyond the atmosphere,[6] the major question in the field concerned their nature. Were they, as Robert A. Millikan of the California Institute of Technology argued, high-energy photons like the penetrating gamma radiation emitted by radioactive substances, or were they charged particles of matter? In theory, a test for determining the answer was simple. If charged (and hence, material), the radiations would be deflected by the Earth's magnetic field. Near the north magnetic pole where the field lines converge, the deflection would be greatest, and near the equator, it would be minimal. In the late 1920s, Millikan and others attempted to discover this 'latitude effect' by measuring cosmic ray intensities at different locations on the Earth's surface. The results of several experiments, however, were equivocal.[7]

Another way to determine the charged nature of cosmic rays consisted of detecting the 'east–west effect.' If positively charged, cosmic ray particles would be deflected by the Earth's magnetic field and arrive in greater quantities from the west. Negatively charged particles would have higher intensities coming from the east. The effect would not only tell whether cosmic rays were charged; it would also indicate which particles predominated by noting a difference in their directional preference.

To search for the effect, Bruno B. Rossi, a physicist at the University of Bologna in Italy, invented a cosmic ray 'telescope' in 1931. The major component of this instrument was a gas-filled tube that had a thin wire stretched along its axis. As demonstrated by the device's inventors, the German physicists Hans Geiger and Walther Muller, the gas would remain barely stable when a negative potential of greater than 2000 volts was maintained on the tube's walls. If a penetrating photon or particle entered the tube, it would ionize a gas molecule and trigger a cascade of electrons, producing a pulse of current that could be amplified and recorded by a mechanical register. To make his telescope, Rossi placed two of these 'Geiger' (or 'Geiger–Muller') counters in a line and connected them in a single circuit so that when a cosmic ray traversed both the tubes, an almost simultaneous discharge, or 'coincidence,' would be produced. By increasing the distance between the tubes, he decreased the telescope's field of view, because only rays coming in a narrow cone could produce the telling coincidences.[8]

While theoretically sound, the experiment for detecting the effect was difficult to perform. Only in 1933 did Rossi and other investigators succeed by finding an excess of radiation in the west direction – an indication that cosmic rays reaching the Earth's surface consisted largely of positively charged particles. High-altitude observations conducted from mountain tops and balloons in the same year demonstrated conclusively the existence of the latitude effect. Along with Rossi's observations, these classical experiments by Millikan and Arthur H. Compton of the University of Chicago indicated the material nature of most cosmic radiation.[9] Cosmic ray research attracted physicists because the radiation constituted a natural source of extremely high-energy particles (up to 10^{14} eV) in the days before powerful accelerators. It quickly became clear, however, that many rays observed on the Earth's surface were not the same as those entering the atmosphere. The incident 'primary' cosmic rays, it appeared, interacted with atmospheric matter to create showers of 'secondary' particles. Only a small fraction of the primary rays made it to the surface unscathed. To study the primary radiation, then, scientists needed to raise instruments high into the atmosphere, an activity that became routine with balloons. Data were obtained by converting measurements into electric signals, broadcasting them from the vehicle and receiving them at ground. From this 'telemetered' information, balloon-borne experiments revealed that the primary radiation consisted mostly of protons, along with a few alpha particles, heavy nuclei, and electrons.[10] In making these observations, cosmic ray physicists demonstrated the usefulness of high-altitude vehicles for examining extraterrestrial phenomena.

Ionospheric research

At nearly the same time that physicists began studying cosmic rays, others became aroused about transatlantic radio transmissions. Demonstrated in 1901 by the Italian-born inventor and entrepreneur Guglielmo Marconi, the 2700 km transmission of 'wireless' messages from Cornwall, England to Newfoundland, Canada surprised people because the curvature of the Earth should have prevented the radio waves from traveling in a straight path between the points. Within a year of the first transmission, the American electrical engineer Arthur E. Kennelly interpreted the puzzling phenomenon as the reflection of radio signals from an ionized region in the atmosphere 80 km high and from the ocean below.[11] A few months later, Oliver Heaviside of England independently proposed a similar explanation. As a reward for their efforts, the reflective region became known as the 'Kennelly–Heaviside layer.' In 1929 Robert A. Watson-Watt of Scotland introduced the more commonly used term 'ionosphere.'[12]

Stimulated by these hypotheses of the atmosphere's ionized region, experimental physicists sought to prove its existence. At an altitude of 80 km, however, the stratum could not be studied directly using instruments carried on sounding balloons, which could only reach one-half that altitude. Consequently, scientists improvised indirect observational methods. Pioneers of the most fruitful approach included Edward V. Appleton and M. A. F. Barnett of England, and Gregory Breit and Merle A. Tuve of the United States, who in 1925 transmitted radio pulses from the Earth's surface into the atmosphere and measured the times required for their return. Because the velocity of radio waves through the atmosphere was close to the known speed of light in a vacuum, the scientists easily calculated the distances to the reflecting layer. In determining heights of between 80 and 225 km for the region, the groups essentially confirmed the existence of the ionosphere.[13]

Meanwhile, research on radio reflection by the atmosphere had begun at the Naval Research Laboratory (NRL), an institution that played a major role in the history of X-ray astronomy and whose background deserves discussion. Recognizing the importance of scientific research for making advances in navigation and the construction of new vessels, the US Navy encouraged investigations in practical fields such as astronomy and hydrography. Under the organizational structure imposed by Congress, however, few opportunities existed to pursue basic research that might have proven useful to the service.[14] While World War I raged on in Europe in 1915, the United States' most famous inventor, Thomas Edison, proposed the creation of a naval laboratory as a means to overcome the deficiency. Research on advanced communications techniques would be among the work undertaken.[15] After a delayed Congressional debate, the laboratory began operations in 1923 on the banks of the Potomac River in southwestern Washington, D.C. A naval officer supervised the 24 civilian scientists and technicians who were grouped in the Sound and Radio Divisions. As Edison had hoped, the investigators would perform fundamental research in communications, a field that had become increasingly important to the US Navy during the war.[16]

Edward O. Hulburt initiated ionospheric research at the NRL in 1924. Holding a Ph.D. in physics from the Johns Hopkins University, he joined the laboratory as head of the Heat and Light Division, a new research section that complemented the original Radio and Sound Divisions.[17] Unlike the older divisions, which derived support from other US Navy bureaus in return for research on specific problems,[18] the Physical Optics Division – as the new Division was renamed in 1931[19] – received funding through the laboratory from general allocations provided by Congress. Consequently, Hulburt enjoyed considerable latitude in choosing fields of investigation. In an early effort, he and A. Hoyt Taylor, head of the Radio Division, verified that a layer of electrons in the upper atmosphere caused the reflection of radio waves around the Earth.[20] Continuing experimental research, Hulburt and others discovered that the ionosphere was not simply a single region of dissociated atoms. Instead, they found it stratified into numerous strata at different altitudes: the 'D-region' (50–90 km high); the 'E-region' (90–160 km high); and the 'F-region' (above 160 km). Besides their altitudes, different electron densities characterized each region.[21]

Because the US Navy needed to maintain radio communications throughout the world at all times, it encouraged NRL scientists to continue ionospheric work. In his next series of investigations, Hulburt studied the mechanisms that caused atmospheric ionization and stratification. While performing this work, he extended the scope of ionospheric research to include studies of phenomena beyond the atmosphere. This was because he realized that the ionizing radiation came from the Sun – a fact revealed by the diurnal and seasonal variations of the ionosphere and the disruption of long-distance transmissions following solar flares. Also working on this aspect of ionospheric research, the English theoretician Sidney Chapman made an important contribution in 1931 by hypothesizing that solar radiations of different wavelengths selectively ionized atmospheric gases at various heights, thereby creating the different layers.[22] While most ionosphericists believed that solar ultraviolet radiation caused the creation of the charged strata,[23] Hulburt found the idea inconsistent with his calculations on the E-region.[24] Instead, he suggested in 1938 (almost simultaneously with the Norwegian Lars Vegard) that X-rays from the Sun might cause the ionization.[25]

This was a remarkable proposal, because most scientists assumed that the rate of X-ray emission from the relatively cool solar surface (5800 K) was miniscule. Discoveries made in the 1930s, however, indicated that the Sun's outer atmosphere – the corona – might have extremely high temperatures and might, therefore, be the origin of a significant flux of X-rays. The first evidence for such temperatures came from Walter Grotrian of Potsdam, who in 1931 measured the Doppler broadening of some absorption lines produced in the corona. Coronal electrons, he concluded, must have velocities up to 7.5×10^8 cm s^{-1}, implying a temperature of about 10^6 K.[26] Using his newly invented coronagraph in 1935, Bernard Lyot of France next observed the broadening of coronal emission lines, suggestive of similarly high temperatures in the solar atmosphere.[27] The Swede Bengt Edlen presented the most convincing evidence for high coronal temperatures in 1939, by identifying coronal emission spectra as originating from atoms that had lost from 10 to 15 electrons. Such ionization required potentials in the hundreds of electron volts, which corresponded to temperatures in the millions of degrees.[28] From this series of observations, it appeared that the Sun might be an excellent source for the ionizing X-rays.

High-altitude rocket investigations

Interrupted by the World War II, Hulburt could not pursue his solar X-ray theory. Testing the theory would have been almost impossible anyway, because the atmosphere absorbed the radiation, whether it be X-radiation or ultraviolet light, at heights beyond those accessible to sounding balloons. By the end of the war, however, the experimental situation had changed dramatically. With rockets, developed first for wartime use in Germany and then for scientific research in the United States, Hulburt and other investigators finally obtained new opportunities to conduct high-altitude experiments.

The initial rocket used for ionospheric research was the 'V-2,' the second of the German 'vengeance weapons (Vergeltungswaffe)' (Figure 2). Supported by the military as a means to circumvent the Versailles Treaty

Figure 2. A V-2 rocket shortly after launch from White Sands Proving Ground. A WAC Corporal rocket serves as the upper stage. Courtesy of the National Aeronautics and Space Administration.

restrictions on other types of weapons development, the vehicle was designed and built between 1930 and 1944 by a team headed by Walter R. Dornberger and including rocket enthusiasts such as Werner von Braun.[29] Using alcohol and liquid oxygen as propellants to produce 249 000 newtons of thrust, the 14-meter tall guided missile had a range of between 290 and 340 km and a maximum speed of 5600 km hour^{-1}.[30] Beginning in September 1944, the German army launched about 3700 of these formidable weapons toward targets in London, southern England, and the European continent. Despite the heavy barrage, the V-2 came too late to turn the tide in favor of the Nazis.[31]

In terms of its range and guidance system, the V-2 rocket surpassed any vehicle developed in the United States. To study the detailed operation of the rockets, the US Army quickly evacuated 300 boxcars full of V-2 parts and equipment from the captured underground factory near Nordhausen before the sector went over to Russian control in June 1945.[32] Shipped to New Mexico for testing at the White Sands Proving Ground, the US Army assembled and launched the rockets ostensibly to train personnel in handling and firing the large missiles.[33] But while the military valued the German rockets largely as testing vehicles, it had no intention of arming them with explosives.[34] Instead, it invited scientific groups to supply instruments for placement in the 0.56-cubic-meter large warhead compartment. Though perhaps only a low-priority objective to the US Army, the conversion of the V-2 from a vengeance weapon to a research rocket made possible the first direct measurements of the upper atmosphere and observations of previously hidden phenomena that occur above it.

Even before the US Army made the V-2s available, several scientists at the NRL began thinking about using rockets for high-altitude explorations. In December 1945 they took a first step toward building their own missiles by establishing a new 'Rocket Sonde Research' branch. Although the members of the branch would eventually design a new vehicle known as the 'Viking' rocket, they temporarily put aside construction plans in January 1946 when they learned about the US Army's offer of the German rockets for scientific research.[35] The NRL scientists obtained encouragement for their proposed work from the Office of Naval Research, which began operations in 1946. The office maintained a philosophy, derived from the war experience, that future military strength depended on balanced and often undirected research efforts. Such work would be conducted by a standing 'army' of civilian scientists working at universities and research centers such as the NRL.[36] The office quickly supported high-altitude rocket studies by providing extensive funding, resulting in what one US Navy investigator called 'an excellent research climate' for peacetime activities.[37]

Taking advantage of a V-2, NRL scientists launched their first experiment into the upper atmosphere on 28 June 1946. The rocket carried a variety of instruments in its nose cone, including an ultraviolet spectrograph. Put there at the request of Hulburt, the instrument could have measured atmospheric absorption of solar radiation at different altitudes. Though insensitive to X-rays, the spectrograph would have been able, perhaps, to discredit the ultraviolet theory for the E-region production by demonstrating that the radiation did not get absorbed there and ionize atoms. Unfortunately, the experiment did not support either theory because retrieval of the instrument proved impossible. After rising to an altitude of 108 km, the vehicle plunged to the Earth in streamlined flight and buried itself in an enormous crater. After weeks of excavation, investigators recovered just a handful of deformed pieces of metal.[38]

Rocket scientists quickly devised means to avoid repeating such catastrophes. Instead of placing the spectrograph in the nose cone, they mounted it in the rocket's tail fin, which was severed before reentry. Having lost its streamlined aerodynamic features, the fin would tumble erratically and slowly to the Earth. When flown in this manner on 10 October 1946, the spectrograph recorded the solar spectrum up to an altitude of 88 km, revealing information below the atmospheric cut-off of about 3000 A (Figure 3). Despite the poor resolution of the lines, scientists discovered several unknown Fraunhofer lines and identified them with lines produced in laboratory gases.[39] Marking the first efforts taken to observe hitherto unexplored spectral regions of a celestial body, the spectrograms were later called 'one of the great achievements in the history of observational astronomy' by the Harvard astronomer Leo Goldberg.[40]

Though revealing new information about the Sun, these and later

Figure 3. The first solar ultraviolet spectra recorded above 30 km, taken from a V-2 rocket flown on 10 October 1946 by NRL scientists. Courtesy of the Naval Research Laboratory.

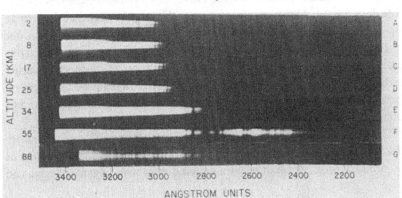

spectrograms did not help verify Hulburt's theory of E-region ionization. Attaining heights sufficient to reduce atmospheric attenuation of the solar radiations constituted only one technical problem to overcome in order to resolve the decade-old debate. Another was photographing the solar spectrum with sufficient exposure times. Because the Sun emits progressively less radiation at shorter wavelengths – the solar spectrum peaks at 4750 Å – photographic films must be exposed for long periods to obtain good spectrograms. This was a difficult task for instruments aboard the spinning V-2 rockets. To be sure, scientists made efforts to develop mechanisms for counteracting the erratic motions of the vehicle and point rocket instruments toward the Sun, but none operated successfully until the early 1950s.

Despite these technological limitations, NRL scientists developed non-spectroscopic means for penetrating the X-ray and ultraviolet wavelength regions shorter than 2000 Å. T. Robert Burnight of NRL's Radio Division invented the earliest system, consisting simply of photographic plates covered by thin aluminum and beryllium windows that transmitted different parts of the X-ray spectrum. Unlike some cosmic rays, which can pass through the steel walls of Geiger counters, X-rays can only penetrate thin metals having low atomic numbers. First flown on a V-2 rocket on 5 August 1948, the plates showed darkening, indicating to Burnight the presence of X-rays in the upper atmosphere (Figure 4).[41] Richard Tousey of the Micron Waves Division adopted a different method for observing radiations at high altitudes. Once a student of Harvard professor Theodore Lyman, Tousey had learned that a phosphor of manganese-activated calcium sulfate absorbed radiation having wavelengths shorter than 1340 Å. Released later by heating, the radiation could be measured by means of a photomultiplier tube.[42] To obtain emissions of restricted wavelength bands with these 'thermoluminescent' materials, Tousey and his colleagues placed suitable filters over the phosphors, which they flew on several rockets launched between November 1948 and February 1950.[43] The data provided the first information in the neighborhood of the Lyman alpha line of hydrogen (at 1216 Å) and other ultraviolet bands. On one of the flights, the phosphors also detected X-rays having wavelengths shorter than 8 Å, thereby providing confirmation of Burnight's earlier finding.[44] Though establishing the existence of X-rays in the atmosphere, these experiments did not demonstrate conclusively that the radiation came from the Sun.[45]

The definitive proof of the solar origin of atmospheric X-rays came from Herbert Friedman (Figure 5), who used rocket-borne Geiger counters as his primary tools. A small, mild-mannered man who began college as an

Figure 4. Blackening of photographic plates, presumably from solar X-rays, from a rocket experiment performed by T. Robert Burnight of the NRL in 1948. Courtesy of the Naval Research Laboratory.

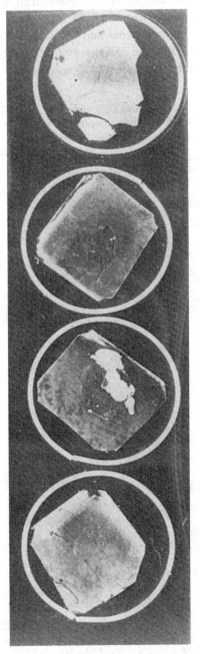

art major, Friedman became a major figure in the history of solar and nonsolar X-ray astronomy. In seeking jobs after his physics education at the Johns Hopkins University, he suffered from anti-Semitism, but with Ph.D. in hand, he obtained a position at the NRL in 1940.[46] Having analyzed X-ray emission lines from elements and alloys as a graduate student,[47] Friedman worked briefly in the Metallurgy Division. In 1941, he transferred to the newly established Electron Optics branch, a unit

Figure 5. Herbert Friedman examining the mirror for an NRL rocket experiment to detect stellar ultraviolet radiations in 1960. Courtesy of the Naval Research Laboratory.

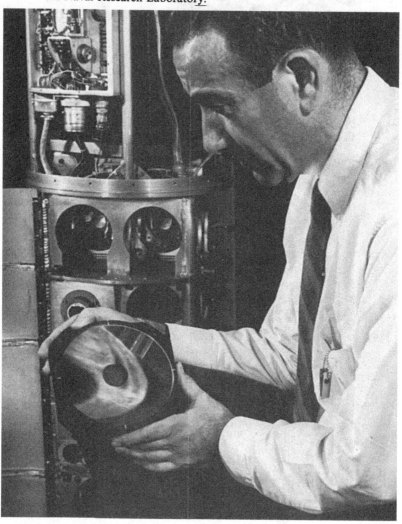

developing means for producing and detecting electromagnetic radiations. While in this branch, which remained under Hulburt's direction in the Physical Optics Division, Friedman developed a technique for orienting and cutting quartz crystals by reflecting X-rays from them.[48] As a reward for inventing this important industrial process used for producing components in radio transmitters, Friedman received the US Navy Distinguished Service Award in 1945.[49] Immediately after World War II, he devised compact photon counters for detecting X-rays and nuclear radiation emitted during atomic bomb explosions.[50]

Working closely with Hulburt, Friedman became interested in using his photon counters to examine solar radiations in the ionosphere. Consequently, he designed a set of detectors having sensitivities in three ultraviolet wavelength regions and one X-ray band.[51] The counters had small windows (about 0.5 cm in diameter) covered by metal films through which radiation passed. Launched by a V-2 rocket on 29 September 1948, Friedman's counters produced data that helped resolve the question of E-region ionization. As the rocket rose, information telemetered from one counter indicated that ultraviolet rays had penetrated to about 70 km – well below where E-region ionization took place. Aside from Hulburt, Vegard, and a few other investigators,[52] most ionosphericists believed the higher E-region absorbed the rays, causing ionization there. This experiment quickly discredited that notion. Between 87 km and the peak altitude of 150 km (in the E-region), however, the counters detected intense X-radiation, whose solar origin was indisputable. Monitoring the rocket's orientation toward the Sun, visible-light photocells confirmed that the Geiger counters pointed toward the Sun when reacting to the radiation. The experiment indicated that solar X-rays were absorbed in the E-region and that they were most likely responsible for the ionization of that stratum.[53] Further experiments led Friedman to conclude confidently in 1952 that solar X-radiation created E-region ionization.[54] Hulburt's theory had finally been verified.

Rocket astronomy

Rocket-borne studies of the ionosphere merely whetted Friedman's appetite for upper atmospheric investigations. During the 1950s he initiated work in two new fields of high-altitude research, the first being solar X-ray astronomy. While other groups, such as those formed at the Johns Hopkins University Applied Physics Laboratory and Princeton University, employed rockets to observe the Sun, they limited their work to the star's ultraviolet radiation, which could be examined using spectroscopic methods. But Friedman's team – including Talbot A. Chubb, a physicist who had worked on electrical discharges,[55] and Edward T. Byram, a mechanical engineer[56] – used nonspectroscopic techniques to study solar X-rays. In most cases the group employed Geiger counters on rockets and satellites.[57] But the US Navy scientists also devised other methods, such as the creative, though simple, technique of photographing the Sun with a pinhole camera and X-ray sensitive plates.[58] In performing any type of research in solar X-ray astronomy, the NRL group remained uncontested until Russian investigators joined the activity in 1959.[59]

Extensive rocket and satellite studies carried out by Friedman's group over a complete solar cycle of 11 years provided the first comprehensive X-ray picture of the nearest star. The dominant X-ray feature was the corona, which radiated uniformly at a temperature of a few million degrees and contributed the bulk of X-radiation in the wavelength region between 8 and 20 Å. During solar maximum, the Sun also produced more energetic emissions, but they originated from solar flares and other localized regions within the outer atmosphere called 'coronal condensations,' where temperatures rose to 100 million degrees.[60] And while the corona and individual spots within it radiated continuous spectra, they also appeared to emit much X-radiation in discrete frequencies that corresponded to excitation levels of highly ionized iron and other elements. Finally, the NRL scientists found that despite the high temperatures involved, the total X-ray intensity of the Sun in wavelengths shorter than 105 Å was extremely small. In each second, the Sun presented only one erg of X-ray energy to each square centimeter of detector window area (written 1 erg cm^{-2} s^{-1}). This quantity contrasted to the solar energy flux in *all* wavelengths of 6.25×10^{10} erg cm^{-2} s^{-1}.[61]

Most of these solar X-ray observations could be readily explained using already-understood physical mechanisms. An important process known to occur, for example, was 'bremsstrahlung,' or 'decelerating radiation.' This type of emission occurred when an ion's electric field altered an electron's direction of motion and kinetic energy. Because energy must be conserved, the interaction resulted in the emission of a photon whose energy reached the electron's kinetic energy as an upper limit. Occurring often in the coronal plasma, these interactions produced most of the Sun's continuous radiation. 'Recombination' emissions also contributed to solar X-radiation. In an isolated case, an ion having a vacant high-energy shell captures a free electron, resulting in the release of an X-ray photon whose energy equals the shell's ionization energy plus the electron's initial kinetic energy. Because the electron's kinetic energy could vary continuously (i.e., not in quantum leaps), the integrated effect of all such interactions in the plasma produced continuous radiation.[62] Lastly, emissions at discrete wavelengths resulted when electrons collided with atoms and excited them into high-energy levels.[63] On returning to their original states, atoms radiated photons. Although all these processes occurred simultaneously, the total amount of solar X-radiation remained small, a fact explained by the low density of the Sun's atmosphere. Because relatively few atoms and ions existed there, these emission processes could not occur as frequently as in a dense plasma.

In addition to solar X-ray astronomy, Friedman's group introduced ultraviolet astronomy as a new specialty by developing techniques in the 1950s for studying stars – not just the Sun – with rockets. A bright source, the Sun was no problem to identify as rocket-borne instruments scanned the body;[64] all other stars had negligible intensities by comparison. But in the night sky, many stars would be scanned, and the identification of particular sources of ultraviolet radiation would be impossible to determine. The exact orientation or 'aspect' of the instruments relative to the celestial sphere, therefore, constituted a major problem to be solved before nighttime rocket astronomy of any sort could proceed.

The solution to the aspect problem was devised for use with what became the workhorse of high-altitude astronomers: the 'Aerobee' rocket.[65] A 'genetic' descendent of the WAC Corporal of World War II fame, the rocket owed its design to James A. Van Allen, a cosmic ray physicist at the University of Iowa. As one substitute for the V-2s, serving double duty for the military and high-altitude researchers, the Aerobee would be used exclusively for scientific research. Measuring six meters long, the rocket weighed only 73 kg – less than an entire payload for a V-2. It burned red fuming nitric acid and aniline without an ignition system ('hypergolicly') in a cleverly designed motor that required no special cooling system. The rocket featured a two-stage booster and sustainer propulsion system that lifted a payload of 70 kg to about 115 km while relying only on its three tail fins for stability.[66] Improved versions of the Aerobee had larger fuel tanks, four fins, and a maximum altitude capability of about 275 km[67] (Figure 6). Remarkably reliable, largely because of its simple design, all the Aerobees had commendable success records of more than 90% from 1948 through 1976.[68] With a low price tag of about $30000 each through the 1960s,[69] the Aerobees afforded scientists on a limited budget the opportunity to repeat a single experiment on different flights or perform a number of original experiments with several rockets.[70]

The feature making the Aerobee amenable to solving the aspect problem was its stability. Flying in the upper atmosphere like a rigid spinning body in a friction-free environment and affected only by a uniform gravitational field, the rocket exhibited simple motions that could be analyzed using

Figure 6. An Aerobee rocket with booster shortly after leaving launch tower. Courtesy of the Naval Research Laboratory.

basic laws of mechanics. Scientists combined this fundamental solution with data provided by a magnetometer, which measured the rocket's orientation relative to the Earth's magnetic field, and information supplied by a sensitive photoelectric system, which identified visible-light signals from bright stars. The result of the analysis described the orientation of the instruments satisfactorily to within one degree of arc.[71] Developed by Friedman's colleagues for the study of nighttime radiations of the upper atmosphere, this system was quickly adopted by rocket astronomers to make ultraviolet surveys of the night sky.

The first success using the new technique came on 17 November 1955, when Friedman's group detected 'plenty of stellar radiation below 1350 Å to permit measurements from rockets with narrow-angle detectors.'[72] Pursuing these results into the 1960s, the group used only Aerobee rockets and gathered information on hot, young stars that emitted most of their radiation in ultraviolet wavelengths.[73] These stars were thought to be the 'predominant steady energy emitters of the galaxies,' and their rate of energy production appeared important for formulating theories of stellar evolution.[74] The team made its most interesting discovery in 1957 when it detected gas clouds obscuring direct radiation from some ultraviolet stars, especially from Spica.[75] Perplexed by this observation because no evidence existed for nebular glows in visible wavelengths, the investigators considered and eliminated several possible explanations. Though totally unexpected, the glows could have demonstrated impressively that condensing gas and dust continuously created new stars.[76] Because of its potential importance, the discovery's confirmation consumed the group's efforts for several years as it repeated the original experiment with highly controlled photon counters. In 1963, the group finally reported that the nebular glows did not exist and were most likely an instrumental aberration on the first experiment.[77] The resolution of the nebular glow problem signaled the end of Friedman's involvement in ultraviolet astronomy.

For two reasons, Friedman's research in the 1940s and 1950s proved extremely important for the future development of X-ray astronomy. Studies of the Sun's X-rays first provided an important description of what scientists assumed to be a typical X-ray emitting star. When they speculated on how intensely other stars might produce X-rays, they first examined the NRL data on solar X-ray emission. As shall be seen in the next chapter, these data did not augur well for the detection of Sun-like bodies emitting in X-ray wavelengths. The NRL studies, therefore, served as a disincentive for earlier work in the field.

Perhaps more importantly, Friedman's work culminated a historical tradition of high-altitude research that made the times ripe – at least from a technological standpoint – for the emergence of X-ray astronomy. He continued the trend, begun in meteorological and cosmic ray research, of using high-flying vehicles to study phenomena *in situ* in the atmosphere and others that occurred far above it. These investigations gave Friedman's group vast experience in developing and flying detection instruments, such as Geiger counters, that would be adopted in the new activity. By making only minor modifications, the early nonsolar X-ray astronomy groups used the same detectors, Aerobee rockets, and orientation techniques that US Navy scientists employed for solar X-ray and stellar ultraviolet astronomy. And although the NRL group did not make the first discovery of a nonsolar X-ray source, it exploited its expertise to react swiftly to the event's announcement in 1962. By designing sophisticated instruments and using them imaginatively, the team quickly became a leader in the new specialty.

2

The political environment

Beyond its technological and scientific heritages, research in X-ray astronomy benefited from a new political environment. Precipitated by the Russian launch of the Sputnik I satellite on 4 October 1957, the new 'order' consisted of increased support for scientific investigations as a way to counter the morale-shattering effect of the Russian achievement. Specifically, Congress established the National Aeronautics and Space Administration, a government agency given a mandate to conduct scientifically sound, yet publicly exciting, explorations of space. The specialty of X-ray astronomy was an outcome – albeit an unintended one – of political policies arising from the Sputnik crisis.

Impact of Sputnik

In the weeks following the orbiting of the first artificial satellite, the American news media reported shock and disappointment. Even though Sputnik's impending launch had been hinted at repeatedly, most observers in the free world did not believe that the Soviet Union could summon the scientific and technological prowess to accomplish such a task.[1] Most stunned and dismayed, of course, were American citizens, who assumed that their country held technical supremacy. For years they accepted the assessment of Vannevar Bush, a major figure in marshalling American scientific efforts in World War II, that technical inquiry in a totalitarian nation could not compare to the untrammeled approach in a free western society.[2] Sputnik shattered that self-confident feeling; some saw it as the 'Pearl Harbor of the Technological War.'[3]

The propaganda effect of Sputnik was tremendous. Launched ostensibly for peaceful scientific purposes during the International Geophysical Year (1 July 1957 to 31 December 1958),[4] the monotonously beeping satellite impressed millions worldwide with its ability to 'outdo the United States in one of the most important scientific fields known to mankind – the frontiers of space.'[5] In six out of seven foreign cities, people told Gallup pollsters that Sputnik dealt a severe blow to the United States' prestige.[6] Moreover, they feared that the perceived scientific and technological advantages held by the Soviet Union would encourage increased neutralism among Third World nations. Or worse, the Soviet Union's achievement might sway unaligned nations, such as the strategic Arab states in the Middle East, to move closer to the Soviet bloc.[7]

Besides shaking international confidence in the United States, Sputnik broadcast a message concerning Soviet military prowess. Though the 84 kg satellite itself did not constitute a weapon, the thrust of its launching rocket suggested the possible superiority of Russian missiles.[8] No longer did American weapons experts discount the claim, made two months earlier, that the Russians had tested powerful intercontinental ballistic missiles with ranges of 8100 km.[9] Equally as important as the rocket's thrust was its guidance system. If the Russians had developed sophisticated techniques for placing a satellite into a tricky orbit, inclined 65 degrees from the equator so that it could pass over the entire world, then maybe they had also developed similar controls for directing missiles accurately onto American targets.[10] Reflecting the fears of many Americans, Democratic US Senator Stuart Symington of Missouri noted that the Sputnik presented 'more proof of growing Communist superiority in the all-important missile field. If the now-known superiority over the United States develops into supremacy, the position of the free world will be critical.'[11]

To placate its shocked citizens, regain its international prestige, and protect itself against a mighty military foe, the United States grabbed for ways to offset the tremendous gains made by the solitary Sputnik. According to *Newsweek* magazine, some Republican politicians suggested that the United States detonate a hydrogen bomb in space so that the world could witness what would supposedly stand as testimony to American technological mastery.[12] And the nation's capital even heard talk of a 'preventative' war with the Soviet Union before it demonstrated its lead in fields other than space exploration.[13] Reacting less severely, many citizens called for a crash program to reduce the 'missile gap.' Harry Stine, a rocket engineer at the Glen Martin missile plant in Denver, remarked bluntly, 'We've got to catch up fast or we're dead.'[14] Meanwhile, President Dwight D. Eisenhower attempted to assuage the public from his retreat in Gettysburg, Pennsylvania, where he remained to play golf.[15] He discounted the Russian success by citing American space achievements since World War II and by describing plans, also linked to the International

Geophysical Year, to launch a satellite with the US Navy's Vanguard rocket.[16] Despite reassurances, members of the administration, the US Department of Defense, and space-oriented professional organizations countered the Sputnik spectacle by proposing a greatly expanded space program.[17]

In the debate occasioned by the Sputnik I launch and intensified in November 1957 by the success of the dog-carrying Sputnik II, two principal foci emerged: the military and science. Behind each banner stood various interested parties. Industrialists rallied around the military, with which they already had enjoyed a profitable association. Moreover, the military had a strong case by arguing that the United States urgently needed to catch up with the Soviet Union's missile capabilities. In addition, it already had experience in directing space research since 1945, and it had produced all the hardware for scientific and defensive purposes. It was the US Army's Jupiter C rocket, for example, that eventually launched the United States' first satellite on 31 January 1958.[18]

Although the military case appeared compelling, other factors augured strongly for a scientific space mission. As argued by members of the National Academy of Sciences, the American Rocket Society, and the Rocket and Satellite Research Panel (the group that coordinated upper-atmospheric research before NASA's creation), the United States first needed to compete in the same race as the Soviet Union to win the propaganda war.[19] The Russians launched the Sputniks for scientific research, and therefore the Americans needed to respond similarly. Also, the American program had to be unclassified because the United States was touted as an open society. Under military control, research results might be suppressed, fostering a sinister image. And as many journalists observed, the military services had shamed the country by allowing the Russians to get into space first.[20] Why reward the military by giving it greater control in the future? Hence, the space program should project a peaceful, scientific complexion and be totally independent of the military establishment, even if much of the technology and research could also serve defensive purposes.

Displeased with the tremendous power held by what he later called the 'military-industrial complex,' President Eisenhower reinforced these views. After consulting members of the Bureau of the Budget and his adviser on science and technology, James R. Killian, Eisenhower quickly sent draft legislation to Congress for creating a new space agency. The proposed civilian-run agency would assume the basic organizational structure of the National Advisory Committee on Aeronautics, a government science body conducting in-house and contracted aeronautical research since 1915,[21] and it would have control over all space projects. Only those programs intimately associated with defensive requirements would be excepted. Approving of the President's move to excise the military from most of the new space program, a *Washington Post* editorialist noted that scientific space ventures might be shortchanged in the hands of the Department of Defense, while missile development would be encouraged.[22] Thus, while Eisenhower recognized the military's needs, he wanted to eliminate the domination already acquired by the services in coordinating space investigations.

Even before the President submitted legislation on 2 April 1958, both the House and Senate established special committees to consider plans for an enlarged space effort. As in the White House, Members of Congress debated the virtues of military and civilian control.[23] Lyndon B. Johnson, chairman of the Senate's Special Committee on Space and Aeronautics, wanted to give the mililtary greater leverage over scientific research than either the President or the House Committee would accept.[24] But the Senate group eventually agreed to allow the new agency and the Department of Defense to decide upon the distribution of space projects without Congressional approval. Where activities held the interests of both the civilian and military sectors, such as in communications and weather forecasting, the agency would be required to coordinate efforts with the services. Consequently, the United States would have two space programs – one civilian and one military – although certain aspects would overlap. With the President's signature on 29 July 1958, the compromise plan became law, and the National Aeronautics and Space Administration began operating the following 1 October.[25]

The legislation creating NASA significantly shaped the character of space research. Most important, according to Homer E. Newell, Jr, NASA's first administrator of such activities, was that a large part of space science had broken out from under the auspices of the military. Now independent of the Department of Defense and its security regulations, space scientists felt less pressure to perform militarily oriented research and could work easily with foreign investigators. The enacting legislation also was noteworthy in its failure to prescribe the specific content of space exploration. Congress merely established the machinery; NASA administrators, responding to advice from independent scientific organizations and researchers, would set goals for investigations as they chose.[26]

Although NASA drew heavily from agencies such as the Naval Research Laboratory for experienced space scientists to serve as administrators[27] – Newell himself came from the naval laboratory – the agency relied on the nation's nongovernment investigators to propose and perform much of the actual space research. But the urgency to explore space did not greatly impress the scientific community as a whole. While it generally approved of the emphasis on technical education and federal support of research sparked by Sputnik – between 1957 and 1961, research expenditures more than doubled to $9 billion (i.e., $9 thousand million) annually[28] – the community feared that the political weight attributed to space might result in overly intensive space programs at the expense of other worthwhile scientific efforts.[29] Some investigators, for example, believed that the large engineering effort required for a manned space program would drain resources from other activities, especially unmanned space science.[30]

Even in disciplines where support for space explorations would seem welcome, active participation by nongovernment scientists did not come immediately. In the astronomy community, for example, few people recognized the potentials for research that the space agency offered.[31] Observations from the unfamiliar environment of space usually required large research teams, development of new electronic equipment, long lead times for projects such as satellite observatories, and no guarantees of success.[32] Classically trained astronomers enjoyed a more conservative, risk-free approach as they worked mostly on an individual basis and with standard tools such as telescopes and spectrographs.[33] And because astronomers also benefited from the new government funding of science that came in the wake of Sputnik, they did not hunger for NASA support. Consequently, members of organizations such as the American Astronomical Society initially showed only tentative interest in space investigations.[34] To stimulate interest in space research by outside scientists, NASA administrators such as Newell found themselves approaching academic and industrial scientists individually and speaking at symposia.[35] In short, NASA had to sell space science.

The stimulus of the National Academy of Sciences

NASA received aid in recruiting scientists and developing research programs from the National Academy of Sciences. The Academy initiated its interest in space exploration in 1952 by coordinating American participation in the International Geophysical Year.[36] With the 'year' ending in December 1958 and the debate raging over the creation of a space agency, Academy President Detlev W. Bronk established the Space Science Board in June 1958 'to survey the scientific problems, opportunities, and implications of man's advance into space.'[37] Keeping with the Academy's tradition of being purely consultative, the Board, chaired by Lloyd V. Berkner, never actually became an operating agency in space research.[38] Instead, it fostered the continuity of American space activities with rockets and satellites and provided advice on space science to NASA, the National Science Foundation, and the Advanced Research Projects Agency of the Department of Defense.[39]

Astronomical investigations in the entire electromagnetic spectrum constituted one research problem conceived of by the Space Science Board. In particular, John A. Simpson, head of the Committee on Physics of Fields and Particles in Space, suggested studies in the previously invisible X-ray and gamma ray wavelengths.[40] Members considered X-ray observations as more than just part of a general program of sky surveys, however. They also saw the investigations as one of the only ways to learn anything about a star's high-energy emissions. As Lawrence Aller, a member of the Committee on Optical and Radio Astronomy, explained, radiation having energies greater than the ionization potential of hydrogen (13.6 eV – the equivalent of light having a wavelength shorter than 912 Å) could not be observed from distant stars. Just as the atmosphere attenuated radiations from beyond it, interstellar space blocked certain spectral regions because of hydrogen absorption. Many stars and nebulae would again be detectable in the X-ray range, in wavelengths shorter than 20 Å,[41] thus making X-ray astronomy a potentially powerful new tool for the discipline.

From a technical standpoint, however, X-ray astronomy appeared to be especially unpromising. Since 1948, solar X-rays had been studied, but the radiation was feeble, constituting only one part in 60 billion of the Sun's total electromagnetic radiation. If scientists assumed that the Sun were a typical radiator, then X-rays from Sun-like bodies at stellar distances would require instruments more sensitive than current detectors by a factor of about a billion![42] Improving the sensitivity of detectors or assuming that other stars might be greater emitters of X-rays would not help much either: quick, 'back-of-the-envelope' calculations showed that the radiation would still be on the fringe of detection. It appeared at first glance that the Sun would remain the sole subject of X-ray astronomy.[43]

Members of the Space Science Board committee were not alone in coming to this conclusion. Almost simultaneously, Louis Henyey and William Grasberger of the Lawrence Radiation Laboratory in Livermore, California performed calculations on possible X-radiation from celestial objects. Carried out in February 1959 for use in a program of observing nuclear explosions in space, the analysis was necessary for detecting detonations above a background of natural X-radiation.[44] The study suggested that X-ray emission would occur only from bodies producing temperatures of the order of one million degrees. Such hot sources might include emission nebulae, young bluish stars of spectral class O or B, the coronas of red supergiants, novae and the Crab Nebula.[45] Even so, quantitative studies of these objects' emissions suggested low fluxes that could never be observed.

Beyond theoretical studies on possible X-ray sources, actual experiments to detect X-ray objects had been performed by Herbert Friedman and the NRL group. Friedman became intrigued by the possibility of nonsolar X-ray sources in July 1956, when he studied X-rays from solar flares.[46] Unexpectedly, his detectors observed some X-rays that appeared to come from somewhere besides the Sun.[47] Stimulated by this hint, he included small X-ray detectors on three rockets along with instruments designed to observe night sky ultraviolet radiation.[48] These attempts to observe X-rays failed, most likely (we now see in retrospect) because the counters were relatively insensitive, having small apertures and being collimated to narrow angles of view. Preoccupied with trying to resolve the knotty problem of the ultraviolet nebular glow around Spica, Friedman did not pursue his hints and speculations much further. Instead, he published an article containing a lower limit for the maximum possible X-ray flux.[49] The report may have suggested to other scientists that nonsolar sources of X-rays would be difficult – and perhaps impossible – to detect.

3

First fruit

Although calculations and a few experimental observations augured poorly for X-ray astronomy, not every scientist discounted the field totally. In fact, several investigators, such as Bruno Rossi, saw enough promise in the field to pursue it. They were aided, of course, by generous government support following the Sputnik crisis and the fact that nature sometimes surprises scientists by revealing unusual phenomena. As a result, Rossi and colleagues at American Science and Engineering, Inc. put nonsolar X-ray astronomy on an observational basis in 1962.

Bruno Rossi and the AS&E, Inc.

By the time Sputnik stimulated space research, Bruno Rossi had already established an enviable career. Born in 1905 and performing pioneering research in cosmic ray physics in the 1920s and 1930s, Rossi distinguished himself by discovering some high-energy particles in cosmic radiation and by measuring the lifetime of the mu-meson. In the course of these investigations he invented basic electronic devices such as the coincidence counters used extensively in cosmic ray and nuclear physics. During World War II, Rossi worked at the Massachusetts Institute of Technology (MIT) Radiation Laboratory and the Los Alamos Scientific Laboratory, where he helped lay the foundation for diagnostic techniques employed by weapons centers. With peace regained, Rossi became a professor at MIT and investigated some newly found cosmic rays and subatomic particles produced in high-voltage accelerators. He also began devising techniques for determining the astrophysical origin of primary cosmic radiation.[1]

Because of his stature as an experimental physicist, Rossi received an invitation to serve as chairman of the Space Science Board Committee on Space Projects, one of the *ad hoc* organizations created by the National

Academy of Sciences in the wake of Sputnik. The committee's main job consisted of defining new research fields that could be investigated in space.[2] An early suggestion coming from Rossi's committee, explicated in a report by Thomas Gold of the Harvard College Observatory, was the study of the interplanetary plasma.[3] Also known as the solar wind, the plasma had been examined only indirectly by measuring the deflection of comet tails; the committee now called for *in situ* measurements of its density, velocity, ionization, magnetic fields, and fluxes.[4] Enthusiastic about the prospects of such studies, Rossi himself initiated the development of a plasma probe at MIT. Funded by NASA, Rossi's team launched an experimental package in 1961 on the Explorer 10 satellite, whose eccentric orbit took it more than 230 000 km from the Earth and provided the first direct evidence of material solar emissions.[5]

The committee also considered X-ray astronomy as a potential problem area. Aware of studies that suggested feeble emanations from sources, Rossi still believed an attempt to look at the X-ray sky would be worthwhile. As a field of unknown potential, X-ray astronomy appealed to Rossi's exploratory instinct – an instinct gained from long experience in cosmic ray research. Reminiscing in 1976, Rossi explained that his earlier investigations:

> had taught me that whenever a scientist ventures into a previously unexplored field, he's likely to be confronted with entirely new and unexpected phenomena. In other words, I did not know anything (or I knew very little) about astrophysics, and at the time when I first became interested in this field I didn't have any specific reason to anticipate the emission of strong fluxes of X-rays from stellar objects. The only thing I knew was that no one had looked for them. Or at least no one had looked with sufficiently sensitive detectors.[6]

Rossi's desire to explore the unknown spectral region of X-rays was not unique, because other Board members obviously believed that such investigations would be worthwhile. Moreover, outside scientists such as Grasberger, Henyey, and Friedman also considered X-ray astronomy. But as his actions relating to the plasma research suggest, Rossi was a determined man who personally assumed responsibility to test his scientific impulses with actual experiments.

Besides being able to express his sentiments to his MIT colleagues, who were now heavily involved in work on cosmic rays, gamma ray astronomy, and the solar wind, Rossi could pursue his instincts through American Science and Engineering (AS&E), Inc., of Cambridge, Massachusetts. Founded in 1958 by two of Rossi's former MIT students – Martin Annis, the company's president, and George W. Clark, a cosmic ray physicist and a new gamma ray astronomer – the firm's staff conducted scientific research and development. During the company's first years, AS&E scientists studied nuclear radiation effects for the Department of Defense and designed instructional science materials for the Physical Sciences Study Committee of Educational Services, Inc.[7] By performing specific tasks under contract, the company's founders hoped to acquire ideas that could be developed into commercial products.[8] Not only the esteemed teacher of AS&E's founders, Rossi was also the firm's principal consultant and Chairman of the Board. These positions gave him a decided impact on the company's research program.

Rossi's belief in the potentials of X-ray astronomy obtained a favorable reception from Riccardo Giacconi, a new member of the AS&E company (Figure 7). Given responsibility over the company's newly expanded

Figure 7. Riccardo Giacconi in 1979 standing in front of an artist's conception of the Uhuru satellite in orbit. Photograph taken by the author.

interest in space science, Giacconi quickly took charge of the X-ray astronomy program. Throughout the next two decades, he would become one of the two leading figures in the field along with Herbert Friedman. Unlike Friedman, Giacconi is a large man who, at least today, exudes self-confidence and the resolve to conduct pioneering research. At the time of the discussion with Rossi in 1959, however, Giacconi was a recent immigrant from Italy whose experience as a scientist had not been particularly satisfying. As he still tells visitors in his heavily accented voice, Giacconi's Ph.D. research at the University of Milan consisted of constructing equipment for detecting esoteric cosmic rays at the top of a mountain. After monitoring the experiment for two years, he had collected data on just 80 cosmic ray events. Because of the scant information, his analysis proved difficult. He dreamt of ways to concentrate more particles on the detector, but nothing ever came of it.[9] Although disappointing, Giacconi's research produced an excellent mental framework for taking in Rossi's ideas. From the start, he considered techniques for concentrating X-rays onto sensitive detectors.

The research program

The first step before making new detectors was to survey available literature and talk to George Clark and other MIT physicists about possible X-ray sources. An internal report resulted from this survey, authored by Giacconi, Clark, and Rossi, called 'A Brief Review of Experimental and Theoretical Progress in X-Ray Astronomy.'[10] Written in late 1959 and prepared for inclusion in proposals to funding agencies in January 1960, the report, summarized below, consisted of answers to two major questions: what astrophysical processes might result in X-ray production? and what possible sources of X-rays might exist?

Speculations about X-ray emission mechanisms centered generally on those known to operate in the Sun, the only confirmed X-ray source. For example, X-rays could be produced from line emission by heavy, multiply ionized atoms similar to those that accounted for some of the observed solar X-ray flux. A second possible process involved bremsstrahlung from hot coronal gases, while a third consisted of recombination emissions. In addition to these three solar processes, the scientists considered another mechanism, 'magnetic bremsstrahlung,' whose contribution to solar X-ray emissions was small, but which might be considerable in other celestial bodies. This process, also known as 'synchrotron' radiation because synchrotron accelerators produce it, occurs when magnetic fields capture and deflect highly energetic electrons. From radio observations the investigators knew that strong fields existed in interstellar space; studies also suggested that radio sources might emit relativistic electrons, thus providing the ingredients for X-ray emission.[11]

Speculation about possible X-ray sources focused on bodies that emitted by the above-mentioned processes. Besides the Sun, which the group slated for more study with small-angle detectors to determine specific X-ray producing regions, the scientists considered four bright types of sources in particular: hot stars such as Sirius; stars exhibiting sudden flare-ups in intensity; rapidly spinning stars having large magnetic fields; and supernova remnants. The last suggestion appealed especially to the scientists, and they paid particular attention to the Crab Nebula, a remnant from a supernova explosion in 1054. As one of the most interesting nearby objects in the sky, the nebula emitted strongly by the synchrotron process in both visible and radio wavelengths. From spectral observations, it appeared that the gas cloud emitted electrons having energies of 10^{12} eV trapped by a magnetic field of 10^{-4} gauss. The AS&E group speculated that more energetic electrons from the Crab might interact with locally stronger magnetic fields, creating X-rays having wavelengths shorter than 20 Å.[12] In making this suggestion, the AS&E group was not unique. As early as 1959, Herbert Friedman of the Naval Research Laboratory noted that the nebula should be one of the most likely X-ray sources in the night sky.[13]

The AS&E survey concluded, therefore, that some celestial sources of X-rays might exist, although their emissions appear too small to detect with state-of-the-art techniques. Placed at a 'close' distance of eight light years, for example, the Sun would present only 10^{-4} X-ray photons per second to a square centimeter of collecting area (written 10^{-4} photons cm^{-2} s^{-1}). Such a flux was about four orders of magnitude too weak for the most sensitive detectors to observe. By assuming the existence of stars brighter than the sun, the scientists still calculated fluxes that were 50–100 times too weak. Though confirming previous notions concerning the practicality (or impractibility) of X-ray observations, Giacconi did not abandon hope after reaching this result. With his experience in cosmic ray physics, he first gave special attention to background cosmic radiation,[14] which could be confused with X-rays in the standard Geiger counter detection system used for solar X-ray studies.[15] By reducing background noise, he thought, the sensitivity of X-ray detectors could be greatly enhanced.

In his search for ways to improve detectors, Giacconi invented a totally novel type of X-ray telescope. From a fortuitous reading of an encyclopedia article on X-ray optics,[16] Giacconi learned of principles developed by Hans Wolter in the late 1940s and 1950s for making X-ray microscopes.[17] The German physicist demonstrated in theory that while X-rays will traverse normal glass lenses without being refracted, they could be reflected off mirrors at 'grazing incidence' – i.e., at very shallow angles of incidence. By using parabolic and hyperbolic mirrors in conjunction, X-rays could be focused onto a suitable detector. Despite this theoretical work, the principle's practical application had been deferred because of difficulties in making mirrors to the small scale required for microscopy.[18] But Giacconi realized that the technological restraints would be eliminated in the construction of *large* mirrors for use in telescopes (Figure 8).[19] Embracing a suggestion offered by Rossi, Giacconi also found that he could nest several mirror surfaces in one telescope, which further increased the instrument's power.[20] Such a telescope would focus X-rays on an improved, but small, Geiger counter. The detector's size assured that the background level of cosmic rays (proportional to the window area of the counter) would remain small, while the telescope's huge collecting surfaces would acquire a large number of X-ray photons. Here then was the design for a big X-ray collecting 'funnel' that could discriminate against unwanted cosmic rays. First estimates suggested that the telescope could detect an X-ray source, similar to the weak sources that had been discussed as possibilities, whose intensity was only 10^{-5} photons cm^{-2} s^{-1}.[21] 'For me

Figure 8. Photograph of X-ray telescope, constructed by the AS&E team in 1966 for solar X-ray observations, resting on a drawing of the mirror's reflecting surfaces. Courtesy of American Science and Engineering, Inc.

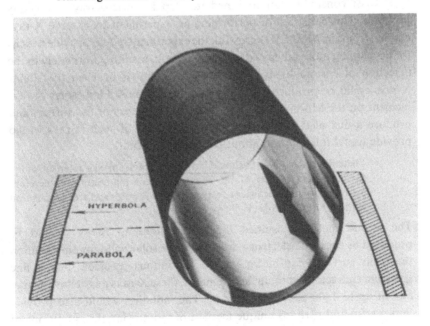

this [telescope] was the key which made X-ray astronomy possible,' reminisced Giacconi. 'Though it would take many years of development, ultimately we could build very powerful instruments to study the radiation from very faint X-ray sources.'[22]

Now enthusiastic about the prospects of his new field, Giacconi attended the first Conference on X-Ray Astronomy, held at the Smithsonian Astrophysical Observatory on 20 May 1960 in Cambridge, Massachusetts. Intended as a 'bull session' (in the words of Albert Baez, chairman of the conference) among a few physicists interested in X-ray instrumentation, the meeting attracted 24 scientists involved in several aspects of X-ray detection. Among them was James Kupperian, a NASA administrator, who recounted attempts made to detect stellar X-rays when he worked with Friedman's NRL group in the 1950s. Paul Kirkpatrick and Albert Baez, two X-ray optics experts, followed with descriptions of techniques used to form X-ray images with reflecting optics.[23] Giacconi and Rossi, meanwhile, described their efforts to design an X-ray telescope.[24] Besides instrumentation, the participants discussed possible sources of nonsolar X-rays. They realized, like the AS&E scientists and others, that the predicted nonsolar X-ray emitters would be weak, but they shared a view that unanticipated sources might be much brighter.

While awaiting the completion of X-ray telescopes, which received initial support from NASA with a grant of about $216 000,[25] Giacconi, Clark, and Rossi considered an intermediate step in their X-ray astronomy program. This consisted of performing an experiment to observe X-rays coming from the Moon. If successful, the experiment would provide a focus for developing more advanced instrumentation, resulting ultimately in the detection of cosmic X-ray sources.[26] The scientists believed that lunar X-rays could be produced through two mechanisms. First, solar X-rays incident on the Moon could cause X-ray fluorescence of the surface and produce a flux of 0.4 photons cm^{-2} s^{-1}. A study of such X-rays would provide useful information about the lunar surface:

> These flourescent X-rays should have a line spectrum characteristic of the chemical composition of the Moon's surface material, and therefore provide a possible means for determining the Moon's surface composition without a lunar probe.[27]

The second method assumed that bremsstrahlung X-rays would be produced as energetic electrons, carried by the solar wind, impinged upon the lunar surface. Exhibiting both a continuous spectrum and a line spectrum characteristic of surface elements, these X-rays might have a large flux of up to 1600 photons cm^{-2} s^{-1}. The radiation would also provide 'a powerful and perhaps unique method of monitoring the conditions of the interplanetary plasma outside the perturbing magnetic field of the Earth.'[28] Of interest not only to X-ray astronomers, the experimental results would aid scientists studying the interplanetary plasma as well. Such a combination of objectives probably reflects the influence of Rossi on the AS&E efforts.

For support in performing this experiment, the AS&E group looked to the Air Force Cambridge Research Laboratories, from which it had previously received contracts.[29] Like the NRL, the research institution, located 20 miles west of Boston, Massachusetts, was concerned with investigations affecting a military service. But unlike its US Navy counterpart, the Cambridge laboratories farmed out work to universities, foundations, and private companies.[30] Sources of funding included the Air Force Office of Aerospace Research, the Air Force Electronic Systems Division, and a number of non-Air Force agencies such as NASA.[31]

Active in space exploration even after NASA took over the national space program in 1958, the US Air Force was receptive to the AS&E proposal to search for lunar X-rays. Through John W. Salisbury, Chief of the Lunar and Planetary Exploration Branch of the Cambridge laboratories, the AS&E scientists obtained support to place a small-aperture (1 cm²) X-ray-sensitive Geiger counter aboard a narrow Nike–Asp sounding rocket.[32] Attempted in 1960, the experiment failed when the rocket engine misfired.[33] After submission of a second proposal for the experiment, the laboratories awarded the company more than $120 000 and provided four larger Aerobee rockets for new attempts.[34] Interestingly, the revised proposal included a new experimental objective. Besides just looking for lunar X-rays, the detectors would search for other sources of nonsolar X-rays. No new evidence suggested that celestial X-rays should be detectable. But aware that its predictions, based on essentially two types of celestial bodies – the Sun and the Crab Nebula – might not have been comprehensive enough to cover all possible cases, the AS&E group maintained that its instruments might still discover an unexpected X-ray emitter.[35]

The next series of experiments employed a detector system considerably more sophisticated than the earlier one. Having available the relatively wide (30 cm diameter) Aerobee rocket, the team developed a larger instrumental payload that included three Geiger counters having window areas of 10 cm² each. The large windows acted as big nets for 'catching' whatever X-ray photons existed, and thereby improved each instrument's sensitivity. To reduce background noise and increase the detector's sensitivity still more, the AS&E group took advantage of its experience in building instruments for cosmic ray research. Behind each Geiger counter, the scientists placed scintillation counters – instruments that produced small flashes of light when energetic particles or photons struck sensitive fluorescent material. Converted to electric pulses by a photomultiplier tube, the scintillations indicated the intensity of the incident cosmic radiation. A cosmic ray particle would now enter the Geiger counter, register a count there, and proceed through the instrument walls until it encountered the scintillation counter, where it would register another count almost simultaneously. Because X-rays could not penetrate to the second instrument, the scientists (or rather, the electronics logic) would discard the counts created by a cosmic ray particle that occurred at about the same time in both detectors. The composite instrument could, therefore, discriminate between cosmic rays and X-rays by detecting the coincidence of events, using the same principles as Rossi's cosmic ray telescope developed in the 1930s. Because it prevented simultaneous occurrences from being interpreted as X-rays, this cosmic ray technique was called 'anticoincidence.' Finally, the scientists broadened the detectors' fields of view (to about 100 degrees in diameter) to observe large portions of the sky during the five minutes above the atmosphere afforded by the rocket. Altogether, the counters' large area and wide field of view, combined with the cosmic ray suppression technique, resulted in an X-ray detector having a sensitivity about 100 times greater than those flown for solar X-ray examinations[36] (Figure 9).

The first attempt to detect nonsolar X-rays occurred on 24 October 1961. With the instruments packed tightly into the nose of the Aerobee, tension mounted among Giacconi and his colleagues in the block house, as it always did before a launch. The blast-off from the long tower at White Sands went flawlessly, and the vehicle reached the required altitude. But as the scientists watched the strip chart for telemetered data, their hopes eroded. Only a straight, flat series of dots appeared indicating that no X-rays had been observed. Soon they learned the reason for the disappointing data: the doors protecting the instruments on the trip up through the atmosphere failed to open. The detectors never even had a chance to look for X-rays![37] Disappointed, the AS&E scientists collected their experiment from the scattered debris on the desert floor and waited for a rescheduled launch date.

On the second try in the late evening of 18 June 1962, almost everything ran smoothly. Attaining an altitude of 225 km, the doors opened properly, and the instruments scanned the sky above the attenuating atmosphere for 350 seconds. Although one Geiger counter failed, the other two operated without problems and provided significant data. In fact the strip recorder revealed that a large intensity of X-rays, having a magnitude of

5 photons cm^{-2} s^{-1} in their peak energy of 4 keV (3 Å), had been detected.[38] The experiment to observe lunar X-rays had been a success!

Or had it? Examining the data closely, the group found that the X-rays came not from the Moon, but from the south azimuthal direction at about

Figure 9. AS&E experiment payload flown on 18 June 1962. Courtesy of American Science and Engineering, Inc.

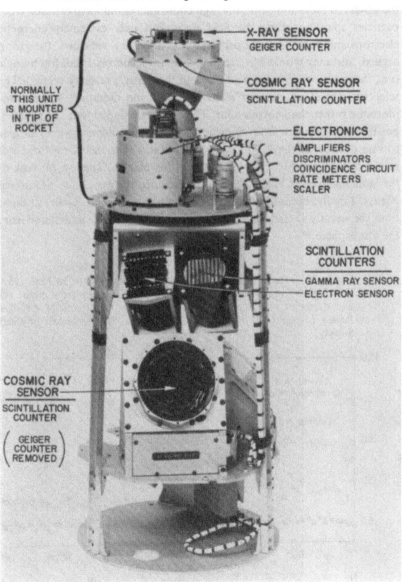

195 degrees (Figure 10), corresponding to the direction toward the center of the Galaxy (Figure 11). Another less intense source also appeared in the east direction at 60 degrees. Meanwhile, diffuse radiation existed everywhere in the background, which was inferred from the fact that counting rates never fell to zero.[39] Considering several explanations for the radiation, the scientists eliminated heavy particle bombardment as a cause. Because the Earth's magnetic field vector had about the same azimuth as the observed radiation peak, the high count rate could have resulted from particles spiraling along the field lines. But only extremely energetic electrons and protons could penetrate the detectors' windows, the group argued, and they would have required an unacceptably small pitch angle (i.e., the angle between the field line and the particle velocity vector). The scientists, therefore, assured themselves that the radiation was electromagnetic rather than corpuscular. They next concluded that the sources' positions, determined from data obtained from an aspect system similar to one devised at the NRL, were correct and that the X-rays did not originate from the Moon or from any of the planets. The low latitude of the launching site at White Sands also excluded the possibility of auroral X-rays. Finally, because of the insignificant level of solar X-rays at flight time, the scientists ruled out the possibility of solar radiation scattered from the atmosphere.[40]

Figure 10. Data returned from the AS&E experiment of 18 June 1962. X-radiation from Scorpius X-1 shows as a large hump from counter 3 and a smaller rise in counter 2. Reprinted by permission of *Physical Review Letters*, Vol. 9, p. 440. Copyright © 1962 American Physical Society.

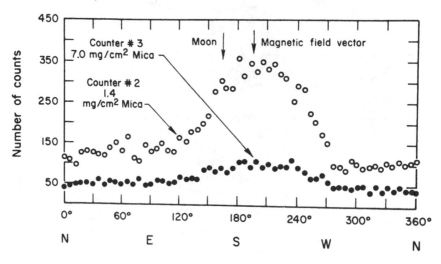

The thorough analysis indicated that the 'Moon' experiment had failed since no lunar X-rays were detected. But intense radiation coming from outside the solar system had been observed, so it must have been emitted by some kind of star or group of stars. But what kind? The group previously predicted that several celestial objects might produce X-rays, but not with the observed intensity. If assumed to be a nearby star, for example, the object would have had to emit X-rays at a rate 10 billion times greater than the Sun.[41] Such a prodigious rate was totally unanticipated. The discovery itself, though, was not completely unforeseen, since the group, inspired by Rossi's exploratory philosophy, held out hope that such an unusual phenomenon might present itself if only someone made the

Figure 11. Chart indicating regions of the sky explored by Geiger counters launched on the AS&E flight of 18 June 1962. The bright X-ray source is located in the shaded box. Reprinted by permission of *Physical Review Letters*, Vol. 9, p. 440. Copyright © 1962 American Physical Society.

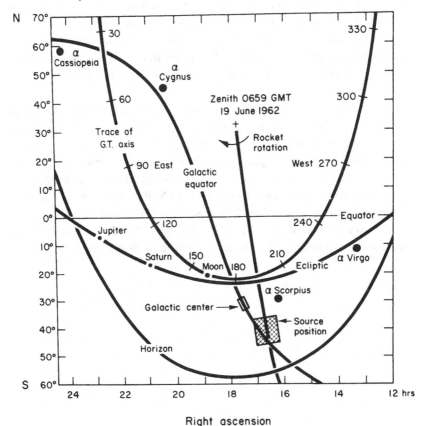

Right ascension

effort to look. Nevertheless, the observation of the strange X-ray emitter, later called Scorpius X-1 (or Sco X-1 for short), excited the scientists tremendously because it revealed a new class of celestial bodies, never conceived of before, that emitted X-rays as their predominant radiation.

According to Herbert Gursky, a low-energy nuclear and cosmic ray physicist who joined the AS&E company in 1961, the Cambridge group was not skeptical of its fantastic results. Proceeding in the traditional manner of well-trained physicists, however, the scientists used the remaining two rockets to confirm the X-ray discoveries.[42] In doing so, they neglected their primary objective of searching for lunar X-rays. But even Salisbury, the group's sponsor at the Cambridge Research Laboratories, did not object to this change of plan, since he too realized that the AS&E discovery held more potential importance than observing X-rays from the Moon.[43] Using similar Geiger counter detection systems, the group launched its next experiments on 12 October 1962 and 10 June 1963. Although the brightest source near the center of the Galaxy hid unobserved below the horizon on the October flight, the instruments detected a source near the galactic plane and another near the Crab Nebula.[44] On the June 1963 flight, the bright source near the galactic center reappeared while another source east of the bright object exhibited itself for the first time.[45] For the AS&E scientists, these investigations satisfactorily verified the existence of celestial X-ray emitters.

The discovery of a bright X-ray source was *not* a serendipitous event, as has been suggested by casual observers of the field's history.[46] While the observation of Sco X-1 was unpredicted, it was not entirely fortuitous either because the AS&E group envisioned the slight possibility of such an achievement. And the AS&E group was not alone in believing this. Almost everyone at the 1960 Conference on X-Ray Astronomy, and even Friedman's group in the late 1950s, had considered sources whose X-ray production mechanisms could not be foreseen. Observing Sco X-1 was the reward nature offered to scientists willing to gamble on a long shot.

SECTION II

THE X-RAY ASTRONOMY COMMUNITY

4

Competition and confirmation

The AS&E results from the June 1962 flight disappointed many scientists who felt beaten to the draw. Though their efforts may not have been as intense as Giacconi's, other investigators initiated work in X-ray astronomy soon after Sputnik's stimulation of space science. But not everyone achieved success even after the main prize was won. Of the two groups sponsored by NASA, one withered away with little trace. The other had a faltering start. Obviously, X-ray astronomy was not going to be an easy science to master. Meanwhile, Friedman and the NRL team quickly challenged the AS&E group for preeminence in the emerging field.

NASA-supported groups

The stimulus for creating the two NASA-supported groups was James E. Kupperian, astronomy branch director of the agency's Goddard Space Flight Center. As a former NRL researcher, Kupperian became interested in X-ray astronomy while working with Friedman in 1956 on the experiment revealing an unconfirmed source of nonsolar X-radiation. In his new position, Kupperian hoped he could entice other scientists to follow up on the NRL X-ray work.[1] During 1959, he contacted several universities and industrial research laboratories seeking scientists interested in performing these investigations with NASA support. Only from the University of Rochester and Lockheed Missiles and Space Company did Kupperian receive favorable responses.

At Rochester, Kupperian discussed X-ray work with Malcolm P. Savedoff, a professor of physics and astronomy. Trained as a classical astronomer, Savedoff already had begun research on detecting gamma rays from celestial sources.[2] He accepted Kupperian's offer to work in X-ray astronomy because he wanted early involvement in the space program, when opportunities were budding. In addition, he expected that detection techniques developed for the new studies could also be used in gamma ray experiments. Receiving a $25000 NASA grant in 1959, Savedoff began constructing rocket-borne detectors that responded to low-energy X-rays.[3] Along with his colleague, Giovanni G. Fazio, a cosmic ray and gamma ray physicist, Savedoff hoped the project would ultimately lead to a satellite observation program. To their dismay, however, the scientists never successfully developed instruments that could withstand the vacuum and vibrations encountered on rocket flights.[4] Work ceased on the project in June 1962 when Fazio left Rochester. No detector was ever delivered to NASA, nor were any results from the program published.[5] As a result, NASA's first attempt to encourage X-ray astronomy proved fruitless.

Kupperian had more luck when he discussed plans with scientists at the research laboratory of the Lockheed Missiles and Space Company in Palo Alto, California.[6] Established in 1955, the laboratory's activities paralleled many of NASA's investigations, although for military reasons. During the late 1950s, the company produced the Polaris submarine-launched missile and the Agena satellite for the US Navy and US Air Force. To solve communications problems during the vehicles' launch and reentry, the company employed scientists to gather information about the upper atmosphere. Other investigators probed the effects of atomic bomb radiations on Polaris missiles. Working on a US Air Force contract, these nuclear scientists fashioned instruments – including X-ray detectors – that could measure the intensity of radiation belts formed during the 1958 atmospheric nuclear weapons tests.[7] Besides contracted work, some physicists undertook undirected research to develop fundamental knowledge that might be useful as the company became involved in nuclear power systems for spacecraft.[8] In summary, the Lockhead laboratory was a place where scientists gained experience in designing high-energy radiation detectors for use in space and where undirected research was encouraged. The environment produced a good source of prospective X-ray astronomers.

Most interested in Kupperian's pitch was Philip C. Fisher, a nuclear physicist hired to design instruments for detecting nuclear radiations in space.[9] Coming from the nuclear weapons center at the Los Alamos Scientific Laboratory, Fisher had gained useful experience while developing methods for obtaining high-energy spectra from fission reactions. An amateur astronomer as well, the Lockheed scientist agreed to search the sky for X-ray emitting sources.[10] Aware of NRL rocket astronomy techniques and previous studies on the possibilities of X-ray astronomy, Fisher was anxious to look specifically at X-ray spectra of sources, from which he hoped to determine 'the abundance of elements in them' and thereby obtain a 'new vantage point from which to study stellar evolution.'[11]

After suggesting a satellite program of research in 1960, Fisher received about $220000 toward a less ambitious plan to observe possible X-ray sources with rocket-borne instruments. Realizing that he needed detectors more sensitive than those used for solar X-ray studies, he developed proportional counters having window areas of between 4.6 and 8 cm². Compared to the apertures of Friedman's solar X-ray instruments, which had areas of about 0.5 cm², these were large windows. To increase the sensitivity of his detectors, Fisher used magnets for sweeping away charged cosmic ray particles that might enter the tubes and be confused as X-rays. Responding to X-ray fluxes of between 10^{-8} and 10^{-9} erg cm^{-2} s^{-1}, Fisher's instruments should have been able to observe sources 10 times dimmer than the upper limit of stellar X-ray intensity reported by Friedman in 1959.[12]

Despite these preparations, Fisher suffered bad luck for the following two years. First came subcontracting delays and NASA red tape that postponed launches from December 1961 and April 1962 until September 1962 for the first one,[13] three months after the AS&E group discovered Scorpius X-1. Thus, Fisher's fire had been stolen. Still, his September launch, and one carried out the following March, could have provided confirmation or discovered other sources. But these hopes also crumbled as the experiments, which recorded several regions of high count rates in the sky, could not identify discrete X-ray sources. The problem could have been caused by X-rays produced as a result of a recent high-altitude nuclear explosion, which dumped high-energy particles into the upper atmosphere.[14] Whatever the cause, Fisher's data never gained much credibility even after publication in the *Astrophysical Journal*. The merits of the results sparked a media debate,[15] and ultimately Fisher admitted that he exaggerated the statistical significance of the high count regions. His conclusions therefore 'must be considered as tentative and suggestive only.'[16]

The experimental failures did not prompt Fisher to bow out of X-ray astronomy research. NASA administrators continued to support him, believing he had acquired valuable experience in constructing X-ray detectors for use in spacecraft that gave him a two year lead on others just beginning X-ray astronomy investigations.[17] Retaining NASA's confidence, Fisher secured funding throughout the 1960s to increase the scientific staff aiding him on new experiments in the field.

The NRL Group

Perhaps most disappointed over the AS&E discovery were Fried-man and his colleagues at the NRL. Many believed that without the circumstances surrounding the Sputnik crisis, they would have discovered Sco X-1 first. After all, Friedman's group had been launching X-ray detectors into space for nighttime experiments since 1956, and it was just a matter of time before one would scan past the bright source. Although small and having narrow angles of view, these instruments were just sensitive enough to detect the emitter, as is now known in retrospect. To improve their chances, the scientists began building larger, more sensitive detectors.[18] If left uncontested, Friedman and his colleagues would have been responsible for opening another field of astronomy.

Because of the AS&E discovery, announced in August 1962 at a Stanford University symposium on X-ray analysis, the NRL group never had a chance to play out this plausible scenario. But, for several reasons, it still wanted to participate in the new field. First, even though the AS&E group labored diligently to demonstrate the nonsolar origin of the X-rays, many scientists would not accept the existence of celestial objects that differed so markedly from the Sun, the only X-ray source previously known.[19] If the emitters existed, the NRL group would confirm them and get credit for good work. Secondly, Friedman recognized that strong X-ray sources might hold tremendous implications for astronomy. Outshining the hot blue stars he studied in ultraviolet wavelengths, X-ray sources might provide new data for constructing models of stellar evolution. He realized too that his group was superbly poised, in terms of building and launching X-ray detectors, to exploit the virgin field.[20] Few other groups in the world had more than a decade's worth of experience in rocket astronomy.

The discovery of an X-ray source spurred Friedman to accelerate his group's construction efforts on an X-ray detector for launch on an already scheduled rocket.[21] Friedman assigned much of the task to C. Stuart Bowyer, who joined the group in late 1962 as a physics graduate student from Catholic University in Washington, D.C. His patented device[22] consisted of a collection of small proportional counters grouped under a sheet of thin beryllium and collimated by a baffle providing a 10 degree field of view. Having an unprecedentedly large window area of 65 cm², the detector was about 10 times more sensitive than the detectors flown by the AS&E group in 1962 and 1963.[23] If the AS&E group had truly observed nonsolar X-rays, then the NRL team would certainly confirm the discovery using this instrument.

Launched in late April 1963, the detector scanned across an intense emitter in the constellation Scorpius. Unlike the AS&E device, which could not accurately locate the source, this instrument's narrow field of view made it possible to determine the object's position above the galactic center at 16 hours 15 minutes right ascension and 15 degrees south declination.[24] Moreover, from a graphical representation of the X-ray intensity, the group could tell that the object was probably a star, though it could have been an extended source (such as a gas cloud) with a diameter of up to five degrees.[25] The detector also scanned past the Crab Nebula, revealing that the nebula emitted X-rays with a flux about seven times lower than that of the Scorpius source.[26] Finally, the instrument observed isotropic radiation from the sky, which might have been of either galactic or extragalactic origin.[27] Comparison of these results with those of the AS&E group showed excellent agreement, although the NRL group provided much more accurate position measurements for Sco X-1 and the source associated with the nebula.[28] Together with the two subsequent AS&E flights, the NRL observation removed any lingering doubts about the existence of tremendously bright celestial X-ray sources.

The independent confirmation of the AS&E discoveries by the highly regarded rocket astronomy group at the NRL set nonsolar X-ray astronomy on a firm observational foundation. The verification was necessary because the conclusion arising from the AS&E observations – that bright X-ray sources of a type not heretofore known exist in space – was so striking and unexpected. While the popular news media may not be considered adequate judges of scientific achievement, it is interesting to note that magazines and newspapers reported the initial AS&E observations only *after* the NRL experiment. *Time* magazine, for example, recounted the Cambridge group's investigations in glowing terms, but only in November 1963,[29] five months after the NRL group first reported its results.[30] *The New York Times*, *Sky and Telescope*, and *Scientific American* followed *Time* with reports of the dramatic discovery of X-ray sources and the birth of a new field of astronomy.[31]

Finally, the NRL confirmation signalled the initiation of a scientific competition with the AS&E group. Throughout the 1960s and early 1970s, the teams competed by designing innovative research programs, inventing new instruments and techniques, and making startling discoveries. Although the scientists at Rochester and Lockheed faltered, Friedman and Giacconi convincingly demonstrated their mastery in the art of X-ray astronomy investigations. Consequently, the leadership and resources flowed largely to these two scientists and their colleagues.

5

'A major area within astronomy'

The discovery of cosmic X-ray sources caused much excitement. While scientists had predicted X-ray emissions from stars, no one imagined that their intensity would be millions of times greater than the Sun's. On one level, then, the discovery excited scientists as do most unexpected findings in any field. But there was more. The existence of the sources suggested that high-energy phenomena occurred more commonly in the universe than previously believed – a view garnered also from observations in radio astronomy. No longer were the transient supernovae, occurring once or twice a century in each Galaxy, the most spectacular cosmic phenomena. Explosively bright X-ray sources lit up the universe as well, but they appeared more enduring than the supernovae. And then there was the omnipresent background radiation, which provoked more questions about the structure and evolution of the universe. In general, the new discoveries implied that the invisible X-ray universe shared few characteristics with its optical counterpart. Thrilled by the prospects of pioneering new avenues in astronomy, scores of scientists entered the young field.

Immediate implications

Although the discovery of the bright X-ray source, Scorpius X-1, appeared to be the most astonishing result of the 1962 AS&E experiment, the observation of a diffuse flux carried more immediate significance for some astronomers. Surprisingly intense, the X-ray background could be used to test the validity of two competing theories dealing with the history of the universe. In one view, articulated by physicist George Gamow, all the matter in the universe originated as a fervid fireball. After exploding in a 'big bang,' matter dispersed into space and then coalesced into conglomerations now observed as galaxies, stars, and other objects. As its most attractive feature, the theory accounted nicely for the observed recession of galaxies.[1] In contrast, the 'steady state' model, developed by cosmologists Herman Bondi, Thomas Gold, and Fred Hoyle, explained the expansion of galaxies without implying the existence of an infinitely dense singularity, as the big bang theory predicted. In this nonevolutionary model, particles of matter maintained a constant density by continuous creation in intergalactic space. Consequently, the universe today is qualitatively no different than in the past, nor any different than it will be in the future.[2]

In early 1963, Hoyle suggested that evidence of the diffuse X-rays observed by the AS&E group in June 1962 buttressed his steady state hypothesis. According to the model, continuously created neutrons would decay into energetic electrons and produce a hot, uniform, intergalactic medium.[3] This hot matter, which had a density that could be computed from the theory, would emit X-rays as bremsstrahlung. From admittedly rough calculations, Hoyle determined that the intensity of this presumed X-radiation agreed with the diffuse flux measured in the rocket experiment.

While Hoyle's calculations supported the steady state cosmology, not everyone believed he performed them carefully enough. Reviewing the computations, Robert J. Gould and Geoffrey R. Burbidge of the University of California at San Diego found that the model predicted an X-ray flux exceeding the observed quantity by a factor of 70. Having greater credibility because they included factors omitted by Hoyle, the new calculations 'must be taken as evidence against the hot universe model.'[4] Although not absolutely discrediting the steady state theory, the initial X-ray observations contributed to the theory's decline in popularity.[5] For the first time, X-ray astronomy experiments had a direct impact on astronomical theory.

Potential implications

Scientists also envisioned great potential implications in the first X-ray data. The reason for this lies partly in the history of radio astronomy – a history familiar to many scientists in the early 1960s. Though Thomas Edison predicted the existence of celestial radio sources in the 1890s, they remained undetected before 1932. In that year, Karl Jansky of the Bell Telephone Laboratories observed radio waves coming from a specific region of the sky, later identified as the galactic center.[6] Work by Grote Reber followed in the late 1930s,[7] but major studies in the field awaited the conclusion of World War II and the use of electronic instruments developed for radar. In 1946, for example, J. S. Hey discovered the first discrete nonsolar radio source, called Cygnus A. Within the next few years, more than 100 radio objects revealed themselves. When optical astronomers searched the sky for these radio sources, however, they rarely found visible objects that could be associated with the low-frequency radiators. Their failure implied that radio astronomers had discovered celestial bodies never before observed. As A.C.B. Lovell of the University of Manchester expressed the widely shared view in 1951:

> [W]e are thus faced with a possibility of the utmost cosmological importance, namely that the Galaxy contains large numbers of stellar objects which are not visible... [T]he origin of these stars seems likely to remain one of the greatest contemporary problems in astrophysics.[8]

The 'invisible' objects of radio astronomy quickly affected the way scientists understood the universe. In particular, practitioners in the new field found space inhabited by more 'violent,' highly energetic objects than previously believed. This new perspective emerged as scientists discovered that most radio signals emitted by discrete sources originated in hitherto unknown astronomical systems that accelerated electrons to nearly the speed of light. Optical astronomers reinforced the view by finally identifying a few sources in visible wavelengths and by examining the spectra of their counterparts. The red-shifted visible spectrum of Cygnus A, for instance, provided a good distance measurement from which its huge luminosity of about 10^{44} erg s^{-1} – a million times greater than the Milky Way Galaxy – could be obtained for the first time.[9] To account for the huge emissions, some investigators postulated that the source consisted of two colliding galaxies, a belief made plausible by its visible appearance.[10] Others explained the energy output by similarly violent causes such as a chain reaction of supernova explosions occurring in the object's core.[11]

The new view of an energetic universe gained further support when radio astronomers discovered quasars in 1960. These star-like objects, whose red-shifted spectra were deciphered in 1963 by Maarten Schmidt of the Hale Observatories,[12] appeared to travel at fantastic speeds – up to 90% of the speed of light – while emitting more than 100 times the energy per second as the entire Milky Way Galaxy.[13] How could these objects produce so much energy in such small volumes? Taking an unconventional approach toward answering this question, some scientists meeting at symposia on relativistic astrophysics in 1963 and 1964 suggested that quasars consisted of supermassive bodies collapsing at their centers; the fantastic power stemmed from gravitational potential energy being converted into electromagnetic radiation.[14]

A year after the first discovery, X-ray astronomy already seemed to mimic radio astronomy's early history. The brilliant source, Sco X-1, resided in a region of the sky where no unusual optical or radio objects appeared. It therefore corresponded to the powerful radio emitter, Cygnus A, an object that had similarly been unobserved in other spectral regions. Perhaps the next stages in the history of radio astronomy would be emulated too. Certainly the initial X-ray observers believed so. Rossi suggested in 1964, for example, that X-ray astronomy investigations might prove as important in changing scientists' understanding of the universe as those made at the other end of the spectrum.[15] During the next few years, as researchers discovered more unusual emitters, Rossi's sentiment gathered more backing. In 1965, the X-ray and gamma ray astronomy group of the National Academy of Sciences' Space Science Board, which included Rossi, Giacconi, Clark, Friedman, and Fisher as members, noted that:

> In three years, the observations [in X-ray astronomy] have progressed from the first evidence of a localized flux to the detection of about a dozen discrete sources. It took radio astronomy a dozen years to progress from Jansky's original discovery to the detection of the radio source Cygnus A... The fact that none of the celestial phenomena known previously had led astrophysicists to predict the existence of X-ray sources even remotely approaching the strength of those observed justifies the expectation that X-ray astronomy will play a fundamental role in advancing our understanding of the universe.[16]

Confidence in the potentials of X-ray astronomy spread quickly. Already in 1964, NASA administrators favorably received a proposal submitted by Giacconi a year earlier for the launch of a small satellite, part of NASA's Explorer series, that would seek new sources.[17] NASA administrators presented the plan to the Congressional oversight committee on space activities in March 1965, arguing that the 'unexpected intensity and character' of the already known sources created the 'urgent requirement' for an all-sky survey using highly sensitive, satellite-borne instruments. With little discussion and no dissent, funding for the program commenced in 1966.[18]

In 1965, NASA planners also began considering an even more ambitious series of X-ray astronomy satellites.[19] The vehicles ultimately became known as High Energy Astronomical Observatories (HEAO), and one of them would carry the powerful AS&E X-ray telescope invented by Giacconi five years earlier. The President's Science Advisory Committee endorsed the basic HEAO concept in 1967, believing that X-ray studies would provide as much information about the cosmos as radio and optical observations.[20] Two years later, NASA's in-house Astronomy Missions Board confirmed the committee's view and rated the HEAO project a top priority.[21] A further recommendation for the project came from the National Academy of Sciences in 1970.[22]

These plans and recommendations provide a good gauge of X-ray astronomy's perceived importance in the 1960s because scientists made them before they understood the physics of X-ray sources. Investigators found that each rocket experiment revealed unusual properties of sources without leading immediately to successful models. A consensus arrived only in 1972 as a consequence of data accumulated by the X-ray Explorer satellite. But the feasibility studies for the HEAO project began well before the revelations of the Explorer. Obviously, then, scientists and administrators showed great enthusiasm for the emerging field. As the Astronomy Missions Board observed in 1969, 'the conditions exist for X-ray astronomy to become a major area within astronomy, with possibly an impact comparable to that of radio astronomy.'[23]

Rapid growth

With such grand expectations, X-ray astronomy attracted an increasingly large community of practitioners who conducted a growing number of investigations. Three indicators document the field's rapid growth. The first concerns 'research activity,' which simply reflects the number of papers written by scientists on X-ray topics. Presumably, if scientists published a large number of articles, they performed a healthy amount of novel work. The second indicator describes the number of people conducting studies in the 'research community' of X-ray astronomers. By definition, a scientist publishing a paper in the field belongs to the community. A final indicator describes 'community growth,' and it provides information on the community's size relative to others in the astronomy discipline.

Research activity subsumes a complex of factors. For example, it is influenced by the size of the research community because more people will probably write a greater number of papers. Alternatively, a special opportunity offered to a single group of experimentalists, such as the chance to fly instruments on a satellite, can affect the amount of research performed without necessarily increasing the community's size. These fortunate scientists will be more active than others and produce more publications. While masking a number of factors, the indicator can be quantified easily by counting the number of articles listed under the heading 'X-Ray and Gamma Ray Astronomy' in the abstracting journal, *Astronomischer Jahresbericht*, and its successor, the *Astronomy and Astrophysics Abstracts*. Table 1 summarizes these data. The most dramatic feature of the data is the growth in numbers of all papers in the field following the discovery of nonsolar X-ray sources in 1962. In that year, the paper announcing the AS&E observations constituted the only one. A year later, six papers hit the press. One of these reported the reflight

Table 1. *Number of papers written in X-ray astronomy 1962–72*

Year	Exp. reports	Theoret. papers	Papers on ground-based observations	Review, popular, & misc. articles	Total papers
1962	1	—	—	—	1
1963	1	5	—	—	6
1964	13	6	—	4	23
1965	13	11	—	7	31
1966	23	15	8	16	62
1967	29	18	20	18	85
1968	42	13	29	14	98
1969	80	39	20	11	150
1970	100	58	22	16	196
1971	114	51	35	12	212
1972	121	53	122	15	311

Data compiled from *Astronomischer Jahresbericht* (Berlin: Verlag W. de Gruyter, published annually between 1962 and 1968); and *Astronomy and Astrophysics Abstracts* (Berlin: Springer-Verlag, published biennially between 1969 and 1972).

Table 2. *Number of American scientists alive in 1975 writing articles in X-ray astronomy 1962–72*

Year	Number of X-ray astronomers	Total number of astronomers	Percentage of X-ray astronomers to total
1962	4	507	0.8
1963	8	571	1.4
1964	7	665	1.1
1965	21	746	2.8
1966	30	878	3.4
1967	45	925	4.9
1968	70	1120	6.3
1969	84	1220	6.9
1970	89	1303	6.8
1971	98	1406	7.0
1972	170	1518	11.2

For important information concerning this table, see note 25, Chapter 5. Unpublished data kindly provided by Thomas F. Gieryn, Department of Sociology, Indiana University, Bloomington, Indiana.

of the AS&E instruments, while the rest, including those written by Hoyle, Gould, and Burbridge, constituted attempts to interpret the first results. In contrast, by the end of 1972, 1175 papers had appeared on topics in the field, of which 311 were contributed in 1972 alone.

Information concerning the research community's size has been provided by Thomas F. Gieryn, a sociologist of science studying the selection of specialties by American astronomers.[24] His data reflect the number of people writing articles in 93 different problem areas of astronomy listed in the same abstracting journals used in making Table 1. Concerned with astronomers in the United States, Gieryn excluded foreign investigators from his data by considering only authors listed in the journals with names included also in the 1975 edition of *Men and Women in American Science* or the 1975 American Astronomical Society membership directory.[25] As

Table 3. *Rank-ordering of 14 astronomical problem areas by number of participants 1961–3 to 1970–2*

Problem area	1961–3 triennia		1970–2 triennia		
	Number of parts.	Rank	Number of parts.	Rank	Change in rank
Radio sources	56	6.5	317	1.0	+5.5
Stellar spectra	64	2.0	259	2.0	0.0
Interstellar space	31	24.0	228	3.0	+21.0
X-ray & gamma ray astronomy	*9*	*69.5*	*222*	*4.0*	*+65.5*
Single, multiple galaxies	73	1.0	208	5.0	−4.0
Stars: physical variables	60	3.0	182	6.0	−3.0
Astronomical accessories	40	15.0	169	7.0	+8.0
Stellar structure & evolution	57	4.5	129	8.0	−3.5
Earth's atmosphere	56	6.5	92	22.0	15.5
Cosmic rays	21	36.5	90	23.5	+13.0
Relativi·y	20	40.5	88	26.0	+14.5
Solar X-rays, UV & gamma radiation	15	57.0	75	28.0	27.0
Crab Nebula	1	90.5	58	37.5	+53.0
Galactic magnetic field	0	93.0	9	79.5	+13.5

Table abstracted from rank-ordering of 93 problem areas of astronomy. Original data kindly provided by Thomas F. Gieryn, Department of Sociology, Indiana University, Bloomington, Indiana.

the data in Table 2 indicate, the number of scientists participating in the field grew dramatically between 1962 and 1972. Beginning with just four scientists in the year of the first X-ray discovery, the research community grew to 170 investigators a decade later. More significantly, the data demonstrate that X-ray research attracted a substantial percentage of American scientists working in astronomical fields. While making up less than 1% of the investigating population in 1962, X-ray astronomers comprised more than 11% in 1972. As one of 93 problem areas of astronomy, X-ray astronomy appealed to a disproportionately large number of scientists.

Gieryn also provides an informative way to demonstrate the community's relative growth by comparing the number of people publishing X-ray studies to those in astronomy's 92 other problem areas. Compiled for three-year periods between 1950 and 1975, these data represent again only American astronomers who were alive in 1975. (Only data relevant to the period between 1961 and 1972 were used.) Each area was assigned a rank based on the number of participants working in it. The area ranked number 1 contained the most investigators writing articles; 93 had the least. Decimal ranks indicate that two or more areas are represented by the same number of scientists. As indicated in Table 3, X-ray astronomy in the 1961–3 period trailed near the end of the list at number 69.5 with only nine investigators. In the 1970–2 triennia, however, only 'Radio sources,' 'Stellar spectra,' and 'Interstellar space' came before it. Now in fourth place, the field claimed 222 participants. Though not the largest area, X-ray astronomy grew more than any astronomy field during this period, jumping 65.5 places in rank. This quantitative measure, and the two discussed earlier, illustrate the dramatic growth in the size of the X-ray astronomy community. If scientists 'voted' for what they considered an important field by conducting research in that field, then X-ray astronomy constituted the rising specialty throughout the 1960s and early 1970s. By these quantitative measures alone, X-ray astronomy stood out as a 'major area' of study.

6

Migrants and money

The rapid growth of the X-ray astronomy community prompts two questions: who were all the new people beginning X-ray investigations? and how did they receive opportunities to perform research? The answer to the first question provides insight on one way modern astronomy has been transformed since World War II. As new fields emerged, participants migrated from external disciplines such as physics. The answer to the second question sheds light on the scientists' intense motivation to enter the specialty, as they exploited traditional as well as innovative avenues to pursue research. Although most investigators received support from regular science organizations, such as NASA, two others obtained assistance in unconventional ways.

Migration into X-ray astronomy

As a new scientific field in the early 1960s, X-ray astronomy could not lure practitioners from existing university graduate programs – the traditional sources for fresh personnel. Instead, it depended on 'migrants' from existing specialties. Because the migrants studied features of celestial phenomena, one might expect them to have come from the astronomy discipline. This expectation, however, does not bear up against data obtained on active American scientists performing experiments in X-ray astronomy between 1962 and 1972: of the 66 experimentalists in this group, 54 (82%) earned their highest degrees in physics, while only 2 (3%) held degrees in astronomy. The remainder had degrees in astrophysics (2, 3%), space science (2, 3%), astrogeophysics (1, 2%), engineering (3, 5%), and geophysics (1, 2%). One scientist, W.B. Smith of the MIT team, had no college degree at all.[1]

The predominance of experimental physicists in X-ray astronomy is easy to understand. With expertise in designing and building instruments that detected high-energy photons or particles, these scientists could apply similar techniques to observe X-rays without much retraining. All they needed to learn was how to build instruments that would operate by remote control and be capable of withstanding the rigors of high-altitude excursions. Few classically trained astronomers had such backgrounds. Instead of developing new instruments themselves, astronomers generally used existing facilities and commercialy available equipment. As described by Nancy Roman, an astronomer and NASA administrator, space scientists with experimental backgrounds were instrument *builders*, while traditional astronomers were instrument *users*.[2] For a similar reason, astronomers generally declined NASA's invitation to perform space experiments after the agency began operations in 1958.

The statistics on experimental physicists in the emerging field gain more meaning on examining the careers of specific investigators. The 'metamorphosis' of William L. Kraushaar, for example, seems to characterize experiences of several other migrants. A cosmic ray physicist, Kraushaar first conducted space experiments in gamma ray astronomy. Interest in the field stemmed from an argument proposed by Philip Morrison of Cornell University in 1957, that gamma rays should be produced in many of the same nuclear and high-energy processes that yielded cosmic rays. By studying gamma rays, Morrison suggested, scientists could learn much about cosmic ray production because, in contrast to material radiation, electromagnetic gamma radiation travels in straight lines to the Earth without being deflected by magnetic fields. Possible sources of large gamma ray fluxes included the radio galaxy M87 and the Crab Nebula, which could be observed by instruments flown for several hours at balloon or satellite altitudes.[3]

Kraushaar began the quest for celestial gamma rays in 1957. Working at MIT with George Clark, he unsuccessfully flew some balloon experiments in search of low-energy (between 1 and 10 MeV) radiation.[4] Immediately after 'these miserable failures,'[5] he devised a scintillation counter for another attempt at discovering celestial gamma rays, this time in the energy range greater than 50 MeV. Able to determine the direction of incident radiation without using a collimation system, the instrument was launched in 1961 on NASA's Explorer 11 satellite. More fruitful than the balloon trials, this experiment detected nonterrestrial gamma rays for the first time. Conflicting with Morrison's predictions, however, Kraushaar and Clark did not find the universe awash with gamma radiation. During the first five months of the satellite's operation, in fact, the MIT physicists detected just 22 celestial gamma ray photons.[6]

Although it provided important information about the density of cosmic rays and electrons in the Galaxy, gamma ray astronomy research proved time-consuming and difficult. Satellite experiments took years to design and fly, resulting in scant, statistically troublesome, and dull data to show for the efforts. In contrast, the X-ray universe looked brighter after 1962, offering enough photons for study in short-lived rocket experiments that usually required less than one year to devise and fly.[7] To many scientists like Kraushaar, who had previous experience in building high-energy radiation detectors, research in X-ray astronomy beckoned. Consequently, when Kraushaar came to the University of Wisconsin in 1965, he undertook, among other projects, a program of rocket-borne X-ray astronomy experiments supported by NASA.[8]

Other physicists shared Kraushaar's positive attitudes concerning X-ray astronomy, but not all had his record of success in space investigations to obtain sufficient funding for a series of rocket experiments. Still, scientists could get an enticing taste of X-ray research at a bargain price by using balloons as launch vehicles. Cheaper than $20000 each for purchase and flight,[9] balloons enabled investigators to try an experiment or two with even less financial risk than with rockets. Balloons presented some restrictions, of course. Cruising at an altitude of about 40 km, instruments could collect only those X-rays unabsorbed by the higher atmosphere, leaving photons having energies greater than 15 keV. Unlike the X-rays first observed from Sco X-1 (having an energy of about 4 keV), these photons could only be detected efficiently with scintillation counters, whose sensitivity at high-energy levels exceeded that of Geiger and proportional counters. Some scientists viewed the limitations as attractions, however, because they had already launched similar instruments on balloons for cosmic and gamma ray studies. Because these physicists required minimal retooling, their migration into the new and promising specialty was remarkably easy.

On 21 July 1964 George Clark demonstrated the feasibility and attractiveness of balloon techniques to potential migrants by examining the high-energy spectrum of the Crab Nebula.[10] Quickly after this exemplary experiment, scientists in the Physics Department at the University of California at San Diego entered X-ray astronomy. As leader of the group, Laurence E. Peterson had already acquired experience in performing balloon-borne gamma ray investigations of the Sun as a graduate student at the University of Minnesota, one of the first institutions involved in such high-altitude research.[11] Because of his background, Peterson's switch to a new line of studies was, like Kraushaar's, relatively simple. In its first experiment in 1965, the San Diego team employed detectors having high-background suppression features to make observations of the Crab in the spectral range between 20 and 250 keV.[12] Likewise, other scientific teams of former gamma ray and cosmic ray physicists, such as those at Rice University, the Goddard Space Flight Center,[13] the Universities of Adelaide and Tasmania in Australia, and the University of Leicester in England, began work in X-ray astronomy by designing balloon-borne experiments.[14]

In many cases, balloon research led to investigations on other vehicles. The San Diego group, for example, won a competition to place a high-energy photon detector, first flown on balloons, on NASA's third Orbiting Solar Observatory in 1967. Though small in size, with a window area of 9.3 cm², the detector[15] was the first to make continuous X-ray observations for more than a few minutes or hours.[16] During a one year period, data transmitted from the instrument established that diffuse X-rays having energies greater than 15 keV were emitted isotropically throughout the celestial sphere.[17] For other groups, such as those at the Universities of Adelaide and Tasmania, balloon X-ray research led to sounding rocket experiments in the late 1960s.

Of course, not all X-ray astronomers were experimental physicists. Theoreticians and ground-based observers contributed to the field's research activity too, as Table 1 shows by organizing data on the number of published papers into four groups: experimental reports; theoretical papers; papers relating to optical, radio, and infrared observations of X-ray sources (i.e., ground-based observations); and review, popular, and miscellaneous articles. The breakdown of the data indicates that not only observations made from space vehicles kept scientists busy. Theoreticians, for example, performed work in the field soon after the AS&E scientists announced their discovery in 1962. Subsequently, as experimentalists made new detections, theoretical astrophysicists followed with interpretations and models. Other than theoreticians, ground-based astronomers contributed by observing X-ray sources in optical, radio, and infrared spectral regions. But these scientists only entered the field after 1966. In that year, as detailed in the next chapter, the AS&E and MIT groups located precisely the position of Sco X-1, enabling terrestrial observers to identify it as a flickering double star system. Soon after the identification, more ground-based observers examined this source and a few more. In turn, these terrestrial observations stimulated speculation on models of stellar systems initially observed in other spectral bands. Similarly, the discovery of pulsing X-ray sources by instruments carried on the AS&E X-ray satellite in the early 1970s roused further ground-based work. In 1972, the number of papers published on terrestrial observations of the objects exceeded those made by X-ray investigators. Never before had this publication phenomenon occurred.

Although important for interpreting data, spurring new interest in the field, and extending observations into different spectral regions, nonexperimental X-ray investigators generally were not active in the field, as indicated by their sparse publications. Theoreticians Hoyle, Gould, Burbidge, and others found that X-ray observations provided useful data for resolving challenging astronomical questions, but they did not publish exclusively on X-ray topics. Likewise, ground-based astronomers investigated interesting problems in X-ray astronomy, but they usually returned to other concerns. The data listed in Appendix 1, which includes the names of the most active authors in the field, support this finding. Of the sample of 126 scientists, only 14 (11%) were theoreticians, while another 14 (11%) were optical, radio, or infrared astronomers. The remaining 98 (78%) conducted experimental X-ray research, illustrating that in terms of numbers alone, experimentalists ruled the new field.

Experimentalists probably dominated the population of the X-ray astronomers for two reasons. The most obvious reflects their greater representation in the scientific community as a whole. A simple statistical study of American Physical Society members shows that 66% published on experimental work in 1970.[18] For theoreticians to have dominated X-ray astronomy in the face of their numerical inferiority would have required them to contribute disproportionately more to one specialty than to others. While scientists in general quickly moved into X-ray astronomy after 1962, as demonstrated in the previous chapter, nothing suggests that the influx occurred selectively among the different types of scientists. Another reason for the experimentalists' majority relates to the instruments they used. With research technology taking months or years to design, scientists developed a commitment to the new field that they were unlikely to abandon after a solitary experiment. They had simply invested too much time and effort. Optical and radio astronomers, on the other hand, made no such commitment when using their usual research tools to observe X-ray objects. And theoreticians had even less at stake. Because of their freedom from all instruments whatsoever (except computers, used for many types of research), they could 'drop in' and 'drop out' at will on specialties such as X-ray astronomy. In short, the nonexperimentalists suffered a high desertion rate, meaning that they worked for relatively brief periods of time on X-ray topics. Unless these scientists were exceptionally productive in that span – and some were – they could not have published the number of papers (at least seven in the sample taken) to be considered significant contributers.

Beyond their background as experimentalists, most X-ray astronomers were also American. In fact, more than 70% of all scientists (experimentalists, theoreticians, and ground-based observers) worked in the

United States (see Appendix 1). Even some foreign scientists such as Minora Oda of Japan conducted early research in the United States with established groups. The heavy involvement of American scientists was probably encouraged by the generous funding, largely by NASA, that nurtured the first X-ray observations. The only country capable of competing with the United States on the same level was the Soviet Union, and from what is known about its space program, experimental activities in a new field such as X-ray astronomy could not be easily initiated. As described by MIT's George Clark, whose professional acquaintances included several Russian physicists and astronomers, Russian scientific research was usually motivated by military needs and planned years earlier by the powerful Academy of Sciences. Even though theoreticians such as Josef S. Shklovsky and V. L. Ginzburg advocated experimental work in X-ray astronomy, the Academy had already committed itself to fields such as cosmic ray physics and remained unreceptive to the new specialty. In a way, Clark pointed out, this inaction was regretable because the Russians could orbit extremely heavy satellite payloads in the 1960s. While Americans launched small instruments on rockets, the Russians could have made tremendous progress with long-lived space observatories. The centralized and inflexible nature of Russian science organization, however, forced the Russians to relinquish a special advantage.[19]

A final observation about migrant scientists concerns their publication habits. Though physicists doing X-ray astronomy needed little retooling for experimental work, they needed to learn how to present their results to the astrophysical community. At first, investigators published their results in the standard physics journals, an action that reflects their disciplinary heritage. But soon enough, as they demonstrated the significance of the observations, astrophysics journals became their primary publication outlets. An examination of publications confirms this contention. During the first four years of experimental X-ray astronomy (1962 through 1965), scientists produced 61 articles scattered among more than a score of journals. Of these, 18 (30%) appeared in physics journals such as the *Physical Review Letters*, which printed the first announcements of the discovery of X-ray sources. Only 11 (18%) showed up in astronomy or astrophysics journals.[20] By 1970, however, the sample of 196 papers included 143 (73%) papers directed toward the astrophysical community, including 40 papers (20%) in the prestigious *Astrophysical Journal* and *Astrophysical Journal Letters*. Only 13 (7%) were found in physics journals.[21]

Opportunities for research

By far, investigators most commonly obtained research opportunities from NASA. Of the 12 American groups performing X-ray astronomy research between 1960 and 1972, all but three received some support from NASA. NASA even funded three of the nine foreign groups (see Appendix 2). NASA provided a positive reception to the new specialty partly because Nancy Roman, head of the NASA's astronomy program, understood the immediate and potential significance of the field's first discoveries. In addition, she and other NASA planners were reminded of X-ray astronomy's importance by various consulting committees, such as the Astronomy Missions Board and advisory groups of the Space Science Board. Although the recommendations of the groups were somewhat self-serving, because the members constituted some of the prospective recipients of the government's largess, NASA stood in a good position to respond favorably to such entreaties and to provide opportunities for examining the X-ray sky.

This favorable disposition resulted primarily from increased funding pumped into the agency after a second major defeat in the space race. Similar in impact to the 1957 Sputnik I launch, the orbiting on 12 April 1961 of the first human, Russian cosmonaut Yuri Gagarin, caused Congress, the President, and much of the country to reassess the space program. Once again, the United States placed a poor second in the competition to prove its prowess in science and technology. Reeling also from the political repercussions of the failed Bay of Pigs invasion attempt on Cuba, which occurred a week later, President John F. Kennedy announced on 25 May a dramatic acceleration of the space program to counteract the deleterious impact of both events.[22] In particular, Kennedy set the goal 'of landing a man on the Moon and returning him safely to the Earth' before the end of the decade. An earlier accomplishment of this goal would have been preferred because officials feared the Russians would attempt a lunar landing to highlight the 50th anniversary celebration in 1967 of the Bolshevik revolution.[23]

Reaffirming the space mission as a major priority, Congress quickly provided NASA with new funds to realize the President's goal. Willing to spend nearly $3 billion for the Apollo Moon project in 1965 and again in 1966, Congress was easily persuaded to allocate about $150 million in each of those years for the physics and astronomy programs. Though relatively small compared with the Apollo budgets, these funds represented an order of magnitude increase over the money dispensed in 1960 (see Table 4). After these peak years, NASA's total budget declined, because the hardware necessary for the lunar project had been largely paid for. Consequently, the allotments for physics and astronomy missions also dropped. Nevertheless, the 1960s was a bountiful decade for funding in the space sciences in general and X-ray astronomy in particular.

Like the circumstances surrounding the Sputnik crisis, X-ray astronomy benefited as an unintended consequence of NASA support for other programs. As a small budget item throughout the 1960s, the field attracted scant attention by Members of Congress or staff attending the all-important authorization hearings of the House Committee on Science and Astronautics. In one sense, X-ray experimenters held an admirable position. Because no public policy directly concerned their research, scientists only had to convince administrators and peer review committees – but not the

Table 4. *NASA appropriations for research and development 1959–72 (in millions of dollars)*

Fiscal year	Total NASA approp.	Total R & D	R & D for Apollo project	R & D physics & astronomy programs
1959	330.9	196.6	10.1	27.6
1960	523.6	347.6	36.1	14.3
1961	964.0	670.4	190.3	44.6
1962	1825.3	1302.5	446.5	88.2
1963	3674.1	2897.9	1160.6	148.6
1964	5100.0	3926.0	2225.0	146.0
1965	5250.0	4363.6	2708.9	160.1
1966	5175.0	4531.0	2971.3	141.1
1967	4968.0	4245.0	2877.9	134.3
1968	4588.9	3925.0	2535.2	145.3
1969	3995.3	3370.3	2025.0	136.9
1970	3696.6	3006.0	1691.1	117.6
1971	3312.6	2565.0	994.5	116.0
1972	3310.0	2522.7	612.2	110.3
1973	3407.7	2600.9	128.7*	156.6
1974	3039.7	2194.0	—	63.6
1975	3226.7	2326.6	—	140.5
1976	3551.8	2677.4	—	162.8
1977	3819.1	2856.4	—	166.3
1978	4063.7	3013.0	—	224.2
1979	4350.2	3292.2	—	285.5

* The Apollo program ended in December 1972 – during the 1973 fiscal year. Source: Jane Van Nimmen, Leonard C. Bruno, and Robert Rosholt, *NASA Historical Data Book, 1958–68*, vol. 1 (Washington, D.C.: NASA, 1976), pp. 115 and 139; and 'NASA Chronological History Fiscal Year Budget Submissions,' for 1969 through 1979, NASA, Washington, D.C.

sometimes poorly informed Members of Congress – about the significance of their work. However, as continued research pointed toward major impacts on astronomy and as investigators and NASA administrators sought more funds for larger experimental packages, House committee members took notice. But such scrutiny began only in the early 1970s and did not appreciably affect the emergence of X-ray astronomy in the previous decade.

Because of X-ray astronomy's inconspicuous profile within the NASA space science program, funding information is difficult to isolate.[24] Nevertheless, studies of NASA contract files provide approximate funding levels for some groups. As might be expected, the AS&E scientists received a large chunk of the X-ray astronomy funding pie: $16 million between 1960 and 1972, excluding the cost of spacecraft.[25] The Lockheed group received about $2 million.[26] And although the NRL scientists obtained funds primarily from the Office of Naval Research, they still received from NASA about $300000 annually from 1967 through the 1970s to pay for rockets, hardware, and salaries for visiting scientists performing research in X-ray astronomy and other space science fields.[27] In total, rough calculations show that X-ray astronomy groups received about $44 million (exclusive of launching vehicles and support services) between 1960 and 1972. This constituted 3% of NASA expenditures for physics and astronomy during the same period. (More complete funding details appear in Appendix 2.)

Unlike scientists who sought funding through NASA, others found research opportunities as part of 'bootleg' operations at the United States' two major nuclear weapons sites. To these scientists at the Lawrence Radiation Laboratory (LRL) in Livermore, California and the Los Alamos Scientific Laboratory in New Mexico, X-ray astronomy studies were secondary activities to the scientists' assigned tasks.[28] At the LRL, X-ray astronomy investigations emerged through the interaction of two semi-independent groups. One was a nuclear weapons testing group, directed by the physicist Frederick D. Seward, which used rocket-borne instruments to measure radiation – especially X-radiation – emitted from explosions in the atmosphere and in space.[29] The second group, led by the nuclear physicist Hans Mark, conducted basic research relating to X-ray emissions from elements bombarded with nuclear by-products. Each unit had special talents and interests in building X-ray detectors: Seward's team built proportional counters sensitive to X-rays having energies less than 3 keV, while Mark's group constructed instruments that measured X-rays in the energy range from 2 to more than 20 keV.[30]

As part of the 1963 Nuclear Test Ban Treaty, the United States established a program of monitoring Russian compliance by launching detection instruments into and above the atmosphere.[31] Contracted by the Atomic Energy Commission, the Sandia Laboratories of New Mexico developed quickly launchable, solid-fueled rockets to carry these detectors.[32] When the monitoring program began in 1963, of course, the AS&E and NRL groups had already reported the discovery of nonsolar X-ray sources. Excited about the field's possibilities, Seward suggested a collaborative effort by the testing and research groups to launch proportional counters on the Sandia rockets, which were being tested without payloads.[33] Because the development of sensitive X-ray detectors was part of the groups' task and since rocket space was available just for the asking, the LRL scientists seized an unusual opportunity to perform X-ray investigations at almost no expense. Unlike university groups, which usually received NASA support, the LRL groups worked on an informal basis with each other and with the Sandia company. Moreover, they never submitted proposals or reports to an arbitrating agency. On the debit side, the high failure rate of the rockets – up to 50% – produced frequent disappointments.[34]

In obtaining a tremendous amount of data from a few important sources, the LRL teams made their most important contributions to X-ray astronomy. Starting with their first experiment in October 1965, the groups repeatedly measured the spectrum of Sco X-1. Using some exceptional instruments, such as a proportional counter having a window area of 2200 cm^2 – an instrument that one member recalled as 'the Mount Palomar [telescope] of rocket-borne X-ray astronomy'[35] – the groups provided information that other scientists gratefully received.[36] And in 1966 the LRL investigators joined with independent optical astronomers in observing the bright object to learn more about its flickering behavior.[37]

At the Los Alamos Scientific Laboratory (LASL) – the first nuclear weapons center, established in 1943 – scientists performed X-ray astronomy research using detectors carried on satellites designed for Project Vela. Managed by the Department of Defense Advanced Research Projects Agency since 1958, the Vela[38] satellites had the task, similar to that of the LRL sounding rockets, of detecting nuclear detonations in the upper atmosphere and in space.[39] To observe X-rays that might have been misidentified as coming from explosions,[40] Jerry P. Conner, a nuclear physicist who directed LASL's high-energy space science group, developed detectors for the twelve Vela satellites launched in pairs between 1963 and 1970.[41] On the fourth pair of vehicles, the scientists made their first attempt at a full-fledged X-ray experiment. Limited by the nature of their mission, these satellites, orbited in 1967, carried small Geiger counters having window areas of 2 cm^2. The spinning vehicles eventually scanned the entire celestial sphere, but they observed only the bright source in Scorpius.[42] Much more significant results came from larger (26 cm² aperture) scintillation detectors carried on the fifth pair of satellites. Soon after launch in May 1969, the detectors observed a variable X-ray source near the boundaries of the constellations Centaurus and Lupus.[43] In more than a month, the counting rate from the source rapidly increased to twice that of Sco X-1 before it declined to zero in September.[44] The object, named Centaurus X-4 (or Cen X-4), constituted the first continuously observed X-ray nova.

X-ray astronomy research at both weapons centers diminished after 1970. At the LRL, investigations waned because underground testing of nuclear weapons proved more profitable in providing data for both the United States and the Soviet Union than originally anticipated. Groups whose major task was to detect clandestine atmospheric explosions therefore became unnecessary. Still, some scientists wanted to continue investigations in the field, and they subsequently left the laboratory. Seward, for example, worked with the X-ray astronomy group at the University of Leicester before joining the Smithsonian Astrophysical Observatory in 1977.[45] The LASL group suspended its X-ray astronomy research too, largely because the weapons center launched no new Vela satellites after 1970. Even though the LASL detectors remained in space throughout the 1970s, instruments launched in late 1970 on the AS&E satellite superceded their ability to measure intensities and spectra of sources.

In summary, the two nuclear weapons laboratory groups took advantage of unusual situations to further X-ray astronomy. Members of the LRL and LASL teams, however, were comprised of experimental physicists having functions to perform other than X-ray astronomy investigations alone. Although the scientists enjoyed opportunities to make good observations of the X-ray sky, they generally regarded themselves as interested observers outside the field.[46] Because of their experimental backgrounds and the limitations imposed upon them by working in high-security weapons centers, the teams could not always pursue interests in the field as did many university groups. Nevertheless, their efforts reveal another aspect of how scientists, intrigued by the prospects of the field after the first observations, carried out investigations in X-ray astronomy.

Seeing how several scientific groups entered X-ray astronomy leads to an observation concerning the wide diversity of experimentalists' research sites. As the classical institution for scientific research, the university is well represented, but interestingly, so are unorthodox sites such as nuclear weapons centers. Moreover, the two major groups, led by Giacconi and

Friedman, worked in commercial and government laboratories. This divergence from traditional patterns should not be considered unusual. Contrary to what might be concluded from these descriptions, the groups' 'marginality' provided no special incentives for pioneering research that were absent in university settings. In fact, as a recent sociological study has pointed out, innovative work was no more likely to be performed in research companies and laboratories than in universities.[47] Perhaps the only grand generalization to make is that the decentralized and pluralistic nature of government support for American science since World War II spawned a variety of research sites, all of which were capable of contributing greatly to the corpus of scientific knowledge.

SECTION III

RESOLVING THE CENTRAL PROBLEM

7

Of mechanisms and a model

The discovery of celestial X-rays in 1962 set the research goal for the rest of the decade. It consisted of answering the central question: what are the physical processes occurring in space that result in copious X-ray production? The unexpected intensity of the emissions implied that scientists were dealing with astrophysical systems having previously unenvisaged characteristics. In approaching an understanding of these systems, some investigators preferred to modify traditional concepts of physics and astronomy, while others considered unorthodox principles. Because of the surprising nature of the first discoveries, the latter approach was never totally dismissed.

From the start, the central question appeared to contain two subsidiary problems. The first concerned the mechanism for converting energy into high-frequency electromagnetic radiation (i.e., X-rays). Not just any conversion process would suffice because candidate mechanisms also had to account for the sources' spectra. Even after succeeding in finding one or more plausible processes, the goal of understanding had not yet been reached. Investigators next needed to solve the problem's second aspect by comprehending the initial energy source. Though scientists understood the nuclear fusion mechanism that produces energy in the cores of stars such as the Sun, a similar process would not do for X-ray objects, which required larger energy stores to account for their greater luminosities. The X-ray source in the Crab Nebula, for example, releases about 7×10^{36} erg s^{-1} in the X-ray portion of the spectrum alone. In contrast, the Sun emits only 4×10^{33} erg s^{-1} in its *entire* spectrum. Astronomers therefore needed new models for stellar energy generation.

Energy conversion mechanisms

Not suprisingly, investigators trying to understand how X-ray sources converted energy into radiation first turned to other fields of astronomy for clues. From radio astronomy, some scientists such as George Clark considered synchrotron radiation, which is produced when energetic radio sources accelerate electrons to relativistic speeds. The charged particles then spiral along magnetic field lines and emit radiation. Given the proper conditions of large magnetic fields and high-energy electrons, X-radiation could in theory be produced.[1] Also considered was the inverse Compton effect, a proposal from theorists James E. Felton and Philip Morrison at Cornell University. In the process, relativistic electrons transfer some of their kinetic energy to photons of starlight during collisions, producing recoil photons of high energy.[2] Conceivably, a supernova remnant radio source could produce the energetic electrons that would collide with photons and yield the observed X-rays. For theoretical and experimental reasons, however, both these processes had to be discounted quickly for Sco X-1 and other sources.[3]

From optical astronomy, other investigators considered blackbody radiation, a process known to be responsible for the energy losses in optical wavelengths from almost all visible stars. The Sun, for example, releases most of its energy from its dense photosphere, which approximates to a blackbody having a temperature of about 6000 K. Scientists were familiar with this emission process and the easily distinguishable spectrum that accompanied it (see Appendix 3 and Table 5). On first glance, however, most scientists discounted the blackbody mechanism too, largely because only dense objects having temperatures in the tens of millions of degrees could produce radiation in large quantities. Although hot and dilute plasmas were known to exist, such as in the solar corona, hot and *dense* bodies were not.[4]

More promising than these commonly known processes was one of novel origin. Proposed in late 1964 by Bruno Rossi, the mechanism consisted of bremsstrahlung in a hot plasma. In this environment, electrons having a wide distribution of kinetic energies move about randomly in a rare atmosphere of ions that is transparent to most radiation (i.e., an 'optically thin' plasma). If the plasma is in thermal equilibrium at a constant temperature, the radiation spectrum from the deflected electrons assumes a familiar Maxwellian (exponential) shape (see Appendix 3 and Table 5). While this specific bremsstrahlung process had been described before 1964 – Friedman's group invoked it in 1960 to explain high-energy X-rays emitted from short-lived solar flares[5] – its efficiency in producing copious radiation was not previously realized and its use for explaining steady cosmic X-ray sources went unnoticed. As Rossi pointed out, the efficiency resulted because fast electrons retain high kinetic energies even after colliding with other rapidly moving particles. The electrons then can radiate strongly when interacting with ions. The situation contrasts one in which energetic electrons encounter slow electrons and protons in a 'cold' plasma, where fast electrons lose much more energy by colliding with other particles than they do by radiation.[6]

Rossi's suggestion of thermal bremsstrahlung had great potential for elucidating the nature of X-ray sources.[7] If the spectrum of a source were identified as exponential, the theory immediately provided information about the plasma temperature and density, two useful parameters for model building. But while investigators quickly identified some objects like Sco X-1 as emitting by the thermal bremsstrahlung process, one major problem remained. If such hot plasma clouds existed and radiated X-rays, what held them together in thermal equilibrium? At temperatures in the millions of degrees, the plasma particles would have tremendous kinetic energies and would dissipate quickly from the X-ray source. Because of this problem, theoreticians initially proposed few models requiring energy conversion by the thermal bremsstrahlung process.[8]

If the developments in X-ray astronomy were to parallel those in the early history of radio astronomy, then the problem of determining how discrete objects generated and emitted so much energy would be a formidable one to solve. Even in the 1950s, radio astronomers only understood the emission mechanisms operating in a few intense objects, and they still had not arrived at any widely shared views on energy sources. The discovery of quasars in 1963 forced new problems as theoreticians considered, unsuccessfully at the time, models of objects generating energy more efficiently than the most efficient nuclear reactions known.[9] Scientists trying to develop models for X-ray sources experienced similar frustrations.

The neutron star model

The effort to develop a neutron star model exemplified the difficulty in finding a solution to X-ray astronomy's central problem. Made in 1964 by Herbert Friedman of the NRL, the attempt testified to scientists' willingness to embrace nontraditional concepts to aid understanding in the field. Although it had existed in publications for 30 years, the neutron star hypothesis had only found a modicum of legitimacy by the 1960s – exactly when people began searching for explanations of X-ray sources.

As a theoretical construct, the neutron star stemmed from attempts made at the beginning of the twentieth century to understand the final stages of stellar evolution. At the time, scientists based models for stellar structure largely on classical theories of thermodynamics and radiation physics: stars appeared to be large gaseous bodies in which gravitational forces held together the hot and ionized matter against the outward pressure of gas and radiation. Knowledge of nuclear reactions for energy production at the core of stars completed the picture of stellar structure, which remains the basis for present theories. However, it is a picture relevant only for stars in 'mid-life,' when the objects occupy positions on the main sequence of the Hertzsprung–Russell diagram.[10] Some stars not appearing on the main sequence, such as white dwarfs, have small sizes (around 10000 km in diameter) and high densities (around 10^5 times greater than the Sun's density). As old, burnt-out stars having undergone gravitational collapse, they posed a dilemma, because from classical physics they should have had high internal temperatures and greater luminosities than observed. Astronomers began understanding the nature of the condensed stars in the 1920s, after Bose, Einstein, Fermi and Dirac described the statistical properties of matter by using quantum mechanics and the Pauli exclusion principle. In dwarfs, electrons have decoupled from nuclei and are as densely packed as permitted by the Pauli principle, in a condition known as degeneracy. Because they rest in their lowest states, the electrons cannot radiate photons, resulting in the stars' low luminosities.[11] With the discovery of the neutron in 1932, George Gamow (and other scientists interested in nuclear energy sources in stars) conceived of similar stellar bodies built up of a degenerate neutron gas. Applying the same basic principles to these bodies, theoreticians developed a model for neutron stars having densities up to 10^{15} times that of the Sun and diameters as small as 10–20 km.[12]

Some scientists found the neutron star concept useful for explaining supernova explosions. Observed since ancient times, supernovae were considered in the nineteenth and early twentieth centuries to be mild stellar flare-ups occurring within galactic nebulae. In the 1920s, however, astronomers determined the extragalactic and distant nature of these nebulae, implying the outbursts were not so gentle after all.[13] Instead they constituted cosmic catastrophes of tremendous proportions.[14] While studying the phenomena, Walter Baade of the Mount Wilson Observatory and Fritz Zwicky of the California Institute of Technology postulated in 1934 that the supernova process represented the transition from a normal star to a neutron star.[15] Supporting this contention, Zwicky pointed out in 1939 that the energy emitted during an explosion was about equal to the energy that would be released when a 'normal' star collapsed from its original 10^6 km diameter to the size of a neutron star.[16]

Interest in neutron stars vacillated for the next two decades. Immediately after the initial studies in the 1930s, neutron star models became less defensible as new work on stellar evolution led most astronomers to conclude that the white dwarf was the inevitable end-point of a star's life.[17] It appeared that a dwarf, which could have a maximum theoretical mass of 1.44 solar masses,[18] would be created after the star exhausted its nuclear energy sources. If the body had a mass greater than the limit, it would eject mass slowly by coronal evaporation or quickly in a supernova explosion.[19] In the late 1950s, however, the outlook for neutron stars changed again, largely because of new studies on stellar nucleosynthesis and the physics of supernova explosions. These investigations, performed by Margaret and Geoffrey Burbidge, William Fowler, and Fred Hoyle suggested that a supernova outburst results from a catastrophic change of state in the core of a highly evolved star.[20] Continuing this line of theoretical research, Alastair G. W. Cameron described a sequence of events in which the degenerate iron core of a star would collapse to a neutron core while the outer layer would be blown away.[21] By reexamining the work performed 20 years earlier by J. Robert Oppenheimer on the relativistic consequences of collapsing stellar matter,[22] Cameron and others devised a variety of models for 'modern' neutron stars.[23]

Neutron stars began tantalizing X-ray astronomers through the work of Hong-Yee Chiu. Having received a Ph.D. from Cornell University in 1959 for his work in particle physics,[24] Chiu examined the effects on collapsed stars of high core temperatures and the dissipation of energy by neutrino production. In the center of a typical neutron star, he argued, the temperature would be about 10^9 K. Energy transmitted from the core by neutrinos and conduction would heat a surface of nondegenerate matter – most likely iron – to a temperature of about 10^7 K.[25] Because of its high density, the iron layer would radiate as a blackbody with a maximum intensity in the X-ray region at about 3 Å.[26] The body would remain in this stage for about 1000 years before its surface would cool to about 10^6 K. At that time, the star would radiate primarily in wavelengths longer than about 30 Å. In optical wavelengths, the star would shine weakly in blue light.[27]

Chiu assembled his work on neutron stars in the monograph, 'Supernovae, Neutrinos, and Neutron Stars,' which circulated as a preprint in 1963.[28] Among its readers was Herbert Friedman, who was attracted to the neutron star hypothesis because it appeared to explain many of the early observations made in X-ray astronomy. In particular, it seemed to account nicely for the enigmatic source in the Crab Nebula, which the NRL group identified in 1963 as a strong X-ray emitter.[29] Even though the gas cloud had been known to emit optical and radio waves by the synchrotron process, the NRL group was puzzled because it believed that the nebular X-rays could not be radiated in the same fashion. The belief stemmed from two bits of information. First, the nebular X-ray flux of about 1.5×10^{-8} erg cm^{-2} s^{-1} greatly exceeded what would be expected from a simple extrapolation of the Crab's known synchrotron spectrum.[30] More importantly, synchrotron radiation of X-rays would require the continuous production at the nebula's center of relativistic electrons having energies as great as 10^{14} eV.[31] This conclusion resulted from calculations of the electrons' lifetimes, which suggested that the charged particles would require regeneration after only a few years in order to produce high-frequency X-rays.[32] But a process for continuously accelerating the electrons was nearly inconceivable at the time.[33]

If one of Chiu's neutron stars existed in the Crab Nebula, however, the problem might be resolved. Because the nebula was known to have resulted from a supernova explosion in 1054, the postulated neutron star would be less than 1000 years old. It would therefore be young and hot enough to emit a substantial flux of X-rays. Intrigued by the possibilities of the hypothesis, Friedman asked Donald C. Morton, a former NRL astronomer then working at Princeton University, to investigate the possibilities of such an X-ray emitting object. Morton concluded, much like Chiu, that a neutron star would probably emit X-rays. Furthermore, a neutron star placed at the distance of the Crab Nebula would account well for the observed flux.[34]

Proponents of the neutron star hypothesis could gain credence by proving that the X-ray emitter was in fact a small ('point') source rather than an extended object within the nebula. At first, the NRL group suggested a program of satellite observations with a high-resolution X-ray telescope such as the one AS&E was developing. Once located, the source could then be searched for in optical wavelengths with the Mount Palomar 200-inch telescope as a faint, blue star.[35]

Like most satellite experiments, this one would require many years to implement. In early 1964, however, Friedman devised a way to test the neutron star hypothesis immediately with a rocket experiment. Through NRL's radio astronomy branch, Friedman learned that the Crab Nebula would be eclipsed by the Moon in July 1964, an event occurring once every decade.[36] Friedman realized that by using the Moon as a screen to cover the nebula gradually, he could determine the angular size of the X-ray emitting object. 'If a neutron star X-ray source existed in the center of the Crab,' Friedman noted, 'the occultation would be expected to produce an abrupt disappearance of X-rays within a fraction of a second of time. A gradual disappearance of X-ray emission would indicate that the X-ray source was an extended cloud.'[37] While uncommon for rocket astronomers to use, the occultation technique did not overwhelm Friedman, who had successfully employed it in 1958 for determining the Sun's X-ray emission regions during an eclipse.[38]

The potential results of the experiment held great importance. Never had a neutron star been observed, but now X-ray astronomy provided a way for the possible discovery of one, leading to a revision of evolutionary schemes. More immediately, the discovery of a neutron star would give X-ray astronomy an elegant model for explaining the energy source in other emitters, especially Sco X-1. Urged on by Josef S. Shklovsky,[39] the Soviet theoretical astrophysicist who had described the visible synchrotron emission from the Crab, the NRL group prepared eagerly for the experiment.

Although enthusiastic, the team faced serious technical difficulties that had to be resolved in the five-month period before the occultation.[40] Unlike its previous nonsolar ultraviolet and X-ray experiments, the group would no longer be surveying the entire sky searching for whatever sources existed. It wanted instead to observe one portion of the sky continuously for the entire duration of a rocket flight. Thus, the vehicle had to be converted into a stabilized platform for constant pointing toward the Crab. For aimed solar experiments, devices called 'attitude control systems' had been designed in the early 1950s. These electronic and mechanical servomechanisms, constructed by the University of Colorado and the NRL, aimed a rocket's instruments at the Sun by tracking the object with phototubes.[41] For nonsolar rocket astronomy, however, similar devices had not been manufactured at the same time because they had no bright targets such as the Sun on which to lock. Trying to remedy this problem, James Kupperian, one of Friedman's former NRL associates and NASA's advocate of X-ray astronomy, requested that the engineering sections of the Goddard Space Flight Center design an attitude control system to suit the requirements of nighttime rocket astronomers.[42] Developed both in-house and under contract to the Space General Company (manufacturer of Aerobee rockets), crude control units used excess gas from the rocket's pressure tanks to despin, orient, and continually align instruments toward chosen celestial objects.[43] Though launched as early as 1959 to provide control within two or three degrees of accuracy,[44] attitude control systems in 1964 still lacked the reliability required by most experimenters: of 14 systems tested in flight by the Goddard center, only three operated as expected.[45] Despite the poor record, the NRL group accepted the risks and obtained an attitude control system for its experiment.[46]

Beyond this technological problem, the team faced the difficult task of launching the rocket precisely at the right moment. Timing was crucial so that the detectors would be in position above the atmosphere to coincide with the occultation of the central part of the nebula, where NRL scientists expected to find the neutron star.[47] The liquid-fueled Aerobee rocket, however, required a lengthy preparation and countdown, and had rarely been launched exactly as planned.[48] Could precise ground crew coordination for the experiment be orchestrated?

Constructing X-ray detectors in the short time available before the launch loomed as the final hurdle. Nearing completion at the laboratory were proportional counters packaged as shallow rectangular boxes covered with mylar windows having surface areas of 114 cm².[49] By operating the counters at high voltages, the group converted the instruments into Geiger counters. Although less sophisticated than proportional counters – being able to record only the number of photons instead of their number *and* energies – the improvised Geiger counters required fewer electronic components. They could therefore be hastily constructed to meet the deadline.[50]

Despite potential problems, the experiment worked perfectly. The take-off on 7 July 1964 occurred precisely on time, and the attitude control system and detectors functioned properly. But while successful in conducting the experiment, the group's hope of substantiating the neutron star hypothesis crumbled. Performing the data reduction immediately – with Friedman uncharacteristically doing much of the tedious work himself[51] – the group discovered that instead of dropping off abruptly as the Moon eclipsed the Crab, the count rate fell gradually. Having an angular width of one arc minute, the X-ray source appeared to have a diameter of about one light year (Figure 12). Such a dimension was 12 orders of magnitude greater than the size of a predicted neutron star.

The absence of a neutron star in the Crab Nebula greatly disappointed Friedman. Enthusiastically adopting the hypothesis, he had colorfully described it as an explanation of the Crab Nebula and other X-ray sources before audiences at Princeton University's Institute for Advanced Studies and the National Academy of Sciences.[52] Nevertheless, the data militated against the interpretation, which received another test two weeks later by George Clark of MIT. Obtaining high-energy spectral data from his initial balloon-borne experiment, Clark found that the neutron star would have needed a surface temperature greater than Morton's theoretical maximum of 16 million degrees.[53]

Meanwhile, the AS&E group demonstrated that the neutron star theory also could not explain the X-ray emission of Sco X-1. By using Geiger and scintillator counters that responded to the wide spectral region between about 1 and 25 keV, the group gathered in August and October 1964 enough data to plot the spectrum of the bright object's radiation. The shape of the spectrum permitted the scientists to conclude that the emitter did not convert energy as a blackbody. Rather, it radiated by either a synchrotron or a thermal bremsstrahlung mechanism.[54] About a year later, the LRL group reported on its first experiment in which it measured the bright source's spectrum with the highest degree of precision yet. With great confidence, the weapons laboratory scientists argued that the

Figure 12. Progress of the occultation of the Crab Nebula measured in seconds of time after launch of the NRL rocket on 7 July 1964. The dashed curves represent the positions of the edge of the moon as it passed over the nebula. A maximum rate of decrease in X-ray flux was observed at about 230 seconds, within the area marked by a circle. The arrow points to the center of the X-ray source, which appeared much larger than a theoretical neutron star. Courtesy of the Naval Research Laboratory.

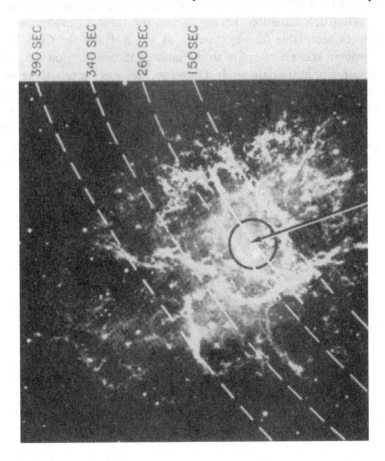

emission resulted from thermal bremsstrahlung in a thin plasma having a temperature of about 60 million degrees.[55] These experiments helped end speculation that Sco X-1 was a neutron star.

In addition to these observations, which discounted the neutron star hypothesis for the Crab Nebula and Scorpius sources, new theoretical studies suggested that neutron stars might never be detected as X-ray emitters.[56] These investigations demonstrated that the cooling rates of neutron stars by neutrino reactions had been underestimated and that the high core temperatures could not maintain themselves for longer than a year, as opposed to Chiu and Morton's calculation of a millennium.[57] In short, both observation and theory conspired against the once attractive hypothesis. Falling into disfavor with the theory was the speculation of blackbody radiation as a plausible energy conversion mechanism for X-ray sources.

Friedman's adoption of the neutron star hypothesis indicated that X-ray investigators were in a quandary concerning the mechanisms and sources of celestial X-radiation. Failing to find the standard models of stellar sources acceptable for objects such as Sco X-1 and the Crab Nebula, Friedman became receptive to the unusual theory that, on the surface, appeared to fit the data nicely. Of course, here was the rub of the matter. In 1964, just two years after X-ray astronomy found an observational basis, experimental data were scarce, resulting from only a few rocket experiments. The groups were obviously dealing with unusual phenomena, and scientists needed information about them to develop complete explanations. If nothing else, the failure of the neutron star theory taught this simple lesson.

8

Research programs

Perhaps sensing the difficulties in trying to resolve the central problem in one fell swoop, Friedman, Giacconi, and their teams established well-defined research programs that did *not* attack the problem directly. Instead, the two groups, which remained the major experimental teams of the decade, developed strategies between 1963 and 1967 that would lead to a general description of the X-ray sky. On the basis of this description, other scientists could perform theoretical work and arrive at the problem's ultimate solution. This approach contrasts one in which the experimentalists' function is to test predictions made by theoreticians, the latter being typical of mature scientific fields that have acquired widely shared conceptual frameworks.[1] X-ray astronomy in the early 1960s, however, had little foundation in either observation or theory. It appeared that detecting more X-ray emitters and studying some of their features would be the best way to attack the central question.

The NRL surveying program

When confronted with enigmatic celestial phenomena, astronomers often resort to searching for more examples of them. They do this because they recognize the difficulties in generalizing from only a handful of examples. By surveying the sky for more sources, astronomers hope to discover clues to their nature from their distribution and intensities.[2] Reasoning in this manner, some radio astronomers in the late 1940s and 1950s approached their central problem by comprehensively surveying the sky. In 1946, only one radio star, in the constellation Cygnus, was known. By 1954, five studies revealed almost 400 discrete radio sources.[3] Performing statistical analyses on the emitters' intensities and distribution, radio astronomers distinguished between different types of sources and learned that some, if not most, radio objects were extragalactic.[4] A year later, the second Cambridge University radio survey revealed nearly 2000 sources. From their distribution and intensities, scientists argued that the universe was evolving, a conclusion that contrasted with the view of an unchanging universe presented in the steady state cosmology.[5] All this had been inferred from scanning work, and it served as a positive example that the NRL group followed.

In choosing a research program consisting of surveying, Friedman's NRL team also took into account its technical background. Since 1955, the group had used surveying techniques for nonsolar ultraviolet observations. The basic procedure consisted of launching a rocket above the atmosphere and allowing it simply to execute the motions of a rigid spinning body influenced by a gravitational force. Deliberately given a slow spin rate, the well-balanced Aerobee precessed in a large cone, which enabled its instruments to scan almost an entire celestial hemisphere. To provide reference points for the rocket at each moment of its flight, the group used auxiliary systems such as magnetometers, horizon detectors, and optical star sensors. According to NRL's Talbot Chubb, the continued use of this rocket technique for observing the universe in a different spectral region was a natural and effective way to contribute observational data in the new field.[6]

The NRL scientists initiated their X-ray surveying program in 1963, when they confirmed the AS&E discovery of celestial X-ray sources. The program continued in 1964 with two experiments, one on 16 June (three weeks before the Crab Nebula occultation experiment) and another on 25 November. Realizing that the rocket's spinning motion would prevent the instruments from observing many photons from individual sources, the team constructed more sensitive detectors to compensate for the short time that each emitter presented itself. The Aerobee payloads, therefore, consisted of Geiger tubes mounted in front of another set of counters in an anticoincident arrangement for reducing cosmic ray background noise. To further increase sensitivity, the scientists gave the instruments window areas of 906 cm^2 – about 14 times larger than the size of the detectors used on the first NRL X-ray instrument of 1963. Honeycomb collimators limited the field to circles of view of about eight degrees in diameter. When combining data from the two flights, the group discovered eight new objects, which complemented the Scorpius and Crab Nebula sources already known.[7] On later surveying flights, such as one in April 1965, the group employed different gas mixtures and more effective anticoincident shields to make smaller, but more sensitive instruments. These detectors, some of which were half the size of those previously used, led to the discovery of several new emitters. By 1967 the number of known sources stood at more than 30. While the NRL team made most of the discoveries, the AS&E and Lockheed groups detected some too.[8]

Because of the increasing number of known X-ray sources, the NRL group instituted in 1965 a nomenclature that remained essentially the same throughout the 1960s. Each source was denoted 'XR' for '*X-ray*,' and numbered within the constellation where it was found according to its intensity, just as radio sources had been catalogued earlier. Hence, the bright source in Scorpius became known as Scorpius XR-1 or Sco XR-1 for short. Likewise, the Crab Nebula emitter, in the constellation Taurus, was abbreviated to Tau XR-1.[9] The AS&E group employed a similar cataloguing system, with the exception that it denoted the sources by an 'X' rather than 'XR.'[10] In the specialty's early days, the emitters' designations became group trademarks, though the AS&E signature was the more frequently used. When scientists discovered more sources in the 1970s, however, they adopted the radio astronomy classification scheme that named objects by their location on the celestial sphere. For example, Sco X-1 became 4U1617-15 (4th '*U*huru' catalog, right ascension *16* hours *17* minutes, declination −*15* degrees).

Even with a sample of only 30 sources, the NRL group could make

Figure 13. Map of X-ray sources in the energy interval between 1–10 keV. Relative intensities of sources are indicated by the sizes of circles. Courtesy of the National Aeronautics and Space Administration.

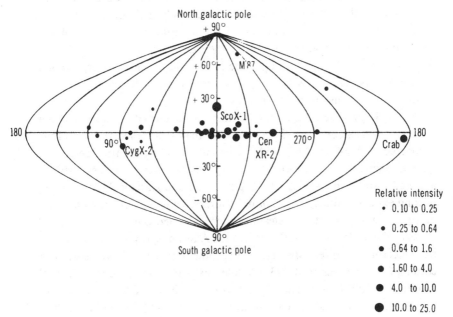

inferences concerning their location in space, number, and luminosities. Most emitters seemed to cluster in two groups close to the galactic plane (Figure 13). One lay in the Cygnus–Cassiopeia region, suggesting that it resided in the same spiral arm of the Galaxy as the Sun, and at an average distance of 1250 parsec.[11] The other association existed in the direction of the galactic center and possibly lay in the Sagittarius spiral arm, at a distance of 2500 parsec.[12] Using 2000 parsec as the objects' average distance and assuming that our Galaxy contains a uniform, flat, disc-shaped distribution of stars 15 000 parsec in radius, the group concluded that it had observed only 2% of all galactic X-ray sources. Hence, it estimated that a total of 1250 X-ray objects inhabited the Milky Way (25 known galactic sources times 50 equals 1250). Based on the presumed power of 10^{36} erg s^{-1} for the known emitters, which they determined by considering observed intensities and estimated distances, the NRL scientists calculated that the Galaxy's total X-ray luminosity would be 7×10^{39} erg s^{-1}.[13]

Although most of the sources clustered around the galactic plane, indicating that they belonged to the Milky Way star system, a few objects appeared displaced. This deviation from the normal distribution suggested that these emitters did not lie close to the Earth, but were possibly extragalactic. From an experiment performed in 1965, the LRL group observed two such sources close to Cygnus A and M87. Cygnus A constituted the first observed discrete radio body, and its identification in visible wavelengths revealed its extragalactic nature. M87, an anomalously bright radio galaxy, emits a bright blue jet extending about one kiloparsec from its apparently active nucleus. If the identification of these objects as X-ray emitters could be confirmed, their detection would have marked the discovery of the first extragalactic X-ray sources. Moreover, it would have made available new techniques – those used in X-ray astronomy – to study these unusual radio emitters.

To check these detections, the NRL group departed from its tradition of making large-coverage sky surveys. Instead it adopted the technique, pioneered in 1964 by the Lockheed group, of using an attitude control system to scan slowly across a selected region of the sky.[14] The new method, adopted also by the AS&E group, gathered good statistical data because of long exposure times to individual sources, while still enabling a search in a particular part of the celestial sphere. Carried out on 17 May 1967, the experiment failed to establish Cygnus A as an X-ray source. However, the detectors observed M87 for the second time and discovered X-rays coming from the quasar 3C273.[15] While this latter radio source shone with an intensity 1000 times weaker than Sco X-1, its intrinsic X-ray luminosity, calculated on the basis of its distance from the Earth of 1.5 billion light years, was over a billion times greater than the Crab Nebula. (The nebular X-ray luminosity of 7×10^{36} erg s^{-1} was confidently known because the object's distance was well established.) Such a prodigious energy production rate made 3C273 'the most powerful X-ray emitter thus far.'[16]

Besides suggesting the existence of strong extraglactic objects, the distribution of sources intimated a correlation between X-ray emitters and more familiar objects. Included among the proposed associations was a correspondence of X-ray sources and supernova remnants residing in nearby groups of hot stars.[17] In light of the discovery that the famous Crab Nebula emitted X-rays, the suggestion was tempting. And as several investigators noted, the speculation appeared to have a good basis as a result of statistical studies.[18]

Finally, the various surveys suggested that X-ray sources might have variable intensities. Though the evidence was partly compromised by differences in observation techniques and uncertainties in data reduction, it appeared that emitters such as Cygnus X-1, which had been observed by the NRL and Lockheed groups, decreased in intensity by a factor of three in a period of three months. The NRL team noted that the variability might have been due to the X-ray star being eclipsed by a secondary object in a binary system. Possibly of great theoretical importance toward understanding the nature of X-ray sources, this feature of variability required further studies, which the group argued should be performed continuously from satellite observatories.[19] Even without using these vehicles, the groups at the Lawrence Radiation Laboratory and the Universities of Adelaide and Tasmania found similar evidence of variability in other sources, making it appear that this characteristic was perhaps common to X-ray emitters.[20]

The NRL surveys, along with less extensive work conducted by other groups, provided an excellent basis for further probing of X-ray astronomy's central question. They had shown that a variety of X-ray sources existed, residing mostly within our Galaxy. A few appeared to be associated with visible radio objects, such as the Crab Nebula, M87, and 3C273. In suggesting possible models for these last X-ray bodies, theoreticians could now draw upon data from three spectral ranges. In addition, the surveys demonstrated that the initial promise of X-ray astronomy as an exciting and potentially important specialty was being kept. Almost every time instruments scanned the sky, they revealed new phenomena. The discovery of variable sources, for example, had been unanticipated, but scientists welcomed it because of its possible implications for building models of X-ray emitters.

The AS&E program of angular resolution and optical studies

Like the NRL approach, the AS&E tack for unraveling the mysteries of the X-ray universe had good precedent in the history of radio astronomy. While the Cambridge University radio group developed a catalog of various sources, other teams, such as the one at Manchester University's Jodrell Bank site, sought more specific information about individual sources. By accurately determining the angular sizes and positions of sources, radio scientists made great strides toward understanding them because they could often identify the objects in visible wavelengths. As these objects became amenable to study by more classical methods, optical astronomers next determined some important parameters, such as distances and luminosities.[21] The discovery of quasars by radio telescopes in 1960 exemplifies this progress. At first the sources did not appear so unusual. But after optical identification, quasars revealed their extraordinary intensities and cosmologically significant distances and speeds. On this basis, Giacconi and his AS&E colleagues believed that a program of high angular resolution determinations leading to optical identifications could play a crucial role in learning about the nature of X-ray sources.[22]

In making their first experiments in this program in 1964, the AS&E scientists sought to determine whether Sco X-1 resembled the Crab Nebula. As the only known X-ray sources at the time, their association appealed to the investigators as the simplest, most straightforward hypothesis. In cooperation with George Clark and other MIT scientists, who were then forming an independent X-ray astronomy group, the AS&E participants tried to measure the angular size of the Scorpius source. If the emitter were a supernova remnant, it would appear extended in space like the Crab Nebula, which had a size of about 12 arc minutes. Angular resolution on

Figure 14. The lay-out of a simple X-ray detector and collimator system. The detector's field of view is determined by the length and width of the collimator's baffles. From Longair, *High Energy Astrophysics*, p. 90.

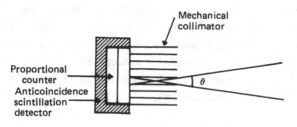

this scale, however, had not yet been attained by conventional collimation techniques. All previous collimators used baffles in front of the X-ray counter, and the size of their field of view determined the angular resolution (Figure 14). For high resolution, scientists needed narrow collimators, but these restricted the number of X-ray photons that could be obtained. In other words, when angular resolution improved, photon counts decreased.[23]

A novel collimator invented by Minoru Oda, a Japanese professor visiting MIT, partially reconciled the conflicting needs of high angular resolution and large numbers of photons. In the words of Paul Gorenstein, a self-styled 'refugee' from experimental physics who helped construct the instrument for the AS&E company in 1966, Oda's collimator was a 'clever idea and extremely simple.'[24] Basically, it consisted of two sets of parallel wire grids that produced shadows on a proportional counter corresponding to the angular size of the source[25] (Figure 15). When used in two rocket experiments in 1964, the collimators enabled AS&E and MIT scientists to define the angular size of Sco X-1 as less than one-half degree and possibly smaller than one-eighth of a degree.[26] But while providing the best angular measurements of the source, the investigators still could not state whether Sco X-1 was a supernova remnant.

The next step in the AS&E program consisted of obtaining better angular resolution measurements of the source and identifying it in optical wavelengths. If shown to have a small angular size, the source was probably a star instead of an extended nebula. Such a measurement would also help in associating the source with a visible counterpart by suggesting to ground-based observers what type of object to search for (i.e., a star or a nebula). Prospects for taking this next step successfully looked good.

Figure 15. A simplified diagram of a modulation collimator. On the left is the shadowing effect of the wires for a point source. On the right, the effect for an object that has an appreciable angular diameter. Courtesy of American Science and Engineering, Inc. as published in *Sky and Telescope*, Vol. 32 (1966), p. 252.

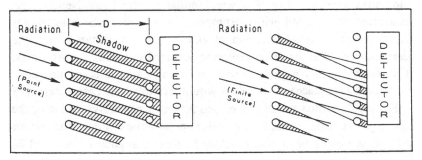

As the brightest X-ray source, Sco X-1 would produce the strongest shadows in a more advanced modulation collimator detection system, making determinations of its angular extent the easiest. And because the source lies far from the galactic center, it would not be obscured by dust, which would render it unobservable in visible wavelengths.[27]

With these considerations in mind, the AS&E group (helped again by some MIT scientists) conceived of an experiment that 'from an instrumentation point of view...represent[ed] a radical departure from previous ones.'[28] The main innovation consisted of using the entire length of an attitude-controlled Aerobee rocket payload section for two large modulation collimators, through which proportional counters viewed Sco X-1. This mode of operation deviated from earlier experiments in which the detectors pointed out of the sides of the vehicle. The long, 64 cm distance between the wire planes and an extra pair of wire grids gave the collimator angular resolution better than one arc minute[29] (Figure 16).

Besides obtaining an accurate measurement of the size of Sco X-1, the collimator system enabled scientists to locate precisely the emitter's position. The key to this accomplishment was making one collimator slightly longer than the other. When slowly scanning past Sco X-1, each instrument modulated the X-ray signals differently, resulting in a kind of vernier effect that scientists used to eliminate possible source locations. As a complementary position measurement system, the group used an aspect camera, which recorded both the star fields that the collimator pointed to and patterns produced by visible light filtering through the grids. When correlated with star fields and X-ray counting rates, these visible patterns described a pair of likely positions for the emitter.[30]

Rivaling the NRL's Crab occultation experiment as one of the most ingenious and well-performed during the decade, the AS&E experiment took place on 8 March 1966 and yielded the desired measurements. It also enabled the AS&E and MIT scientists to make important inferences about Sco X-1.[31] The determination of 20 arc seconds for the bright object's angular size, for example, ruled out the possibility of the emitter being a supernova remnant. For if it were similar to the Crab Nebula, this measurement, coupled with assumptions concerning the rate of gas expansion in a supernova remnant and the object's distance from the Earth, implied that the stellar explosion occurred less than 50 years earlier. This result appeared impossible to accept because such a recent supernova outburst taking place in our Galaxy would certainly have been observed.[32]

Having discarded the theory of Sco X-1 as a supernova remnant, the investigators next tackled two other possibilities. First, the visible counterpart could have been a nebula having an angular diameter of 20 seconds, the maximum size allowed by the experiment. A simple extrapolation of the object's intensity from the X-ray to visible portions of the spectrum suggested that if it were a nebula, the source would have a surface brightness of about +19th magnitude per square second of arc. In independent efforts, Hugh M. Johnson of the Lockheed group and the AS&E-MIT scientists surveyed plates of the region near Sco X-1, but they found no plausible candidates.[33] The second, and now more likely possibility, suggested that the counterpart of the X-ray source was a point emitter. If this conclusion were correct, the object would shine like a +13th magnitude blue star in visible wavelengths.[34]

Figure 16. Schematic diagram of the collimators and counters used by the AS&E-MIT group to determine the angular size and location of Scorpius X-1 on 8 March 1966. Courtesy of American Science and Engineering, Inc. as published in Lodewijk Woltjer, ed., *Galaxies and the Universe* (New York: Columbia University Press, 1968), p. 56.

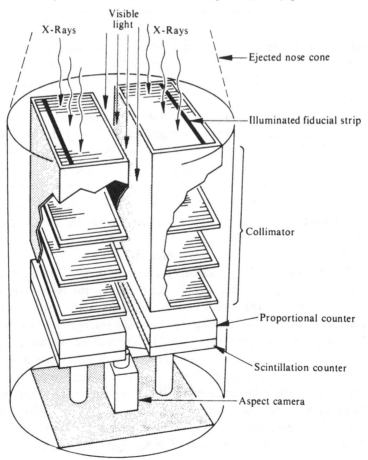

The optical identification of Sco X-1 followed quickly. Unaware of an agreement in which the AS&E scientists would forward position data exclusively to Alan Sandage of the Mount Palomar Observatory, Oda transmitted the information to his colleagues at the University of Tokyo Observatory.[35] On examining photographic plates taken with a 74-inch telescope, the Japanese observers noted a blue star of +12.6th magnitude within a minute of arc from one of the two most likely source positions.[36] Within a week of this detection, observers at Mount Palomar used the 200-inch telescope to verify the identification. It appeared that the brightest X-ray object, discovered four years earlier, had finally been identified with a known emitter in the sky (Figure 17).

Previously an object of no special significance and photographed without comment since 1896,[37] the optical counterpart of Sco X-1 immediately became the focus of intensive ground-based observations. Visible-light studies of the source would be especially important to X-ray astronomy researchers because, as the apparently valid prediction of Sco X-1's visible intensity implied, both the X-ray and optical emissions resulted from the same energy conversion mechanism. In visible light, then, astronomers studied the X-ray production process directly, but without the need to perform space experiments.[38] As a result of these investigations, scientists quickly found that the counterpart appeared to have many features of an old nova, such as a similar emission spectrum and characteristic fluctuations in brightness.[39] Optical identification also enabled scientists to estimate its distance of 250 parsec. With this determination, they next calculated the object's X-ray luminosity to be about 10^{36} erg s^{-1}. In the visible portion of the spectrum, the source emitted only 10^{34} erg s^{-1}.[40] The value for the X-ray luminosity agreed with that estimated earlier by the NRL group from less substantive methods. It also was approximtely equal to the X-ray power of the Crab Nebula. Such consistency must have been comforting.

As the AS&E group had hoped, observations of Sco X-1 by optical (and soon by radio and infrared)[41] astronomers motivated theoretical investigations on the conversion mechanisms and energy sources in discrete X-ray emitters, leading rapidly to the emergence of a new model for X-ray sources. Identifying the counterpart as an old nova provided the major clue, because scientists knew most of these objects as binary systems in which a compact object, such as a white dwarf, was coupled gravitationally to a large and bright star.[42] As early as 1964, the Japanese theorists Satio Hayakawa and Masaru Matsuoka hypothesized that mass ejected by a large star in a similar binary system might produce a shock wave as it hit the smaller star, creating high temperatures and X-ray emissions.[43] Because it was included in a discussion with other hypotheses and lacked observational evidence, however, the binary theory received little attention at the time. But Bruno Rossi's announcement of the identification of Sco X-1 as a possible nova at an international astronomy symposium in August 1966 revived interest in the theory. At the conference itself, Rossi, Burbidge, Shklovsky, Kevin Prendergast, V.L. Ginzburg, and others convened to discuss the merits of the binary model.[44]

Figure 17. On a Mount Palomar photograph, two rectangles (one by two minutes of arc) mark possible positions of Scorpius X-1 as indicated by March 1966 rocket data. Courtesy of American Science and Engineering, Inc.

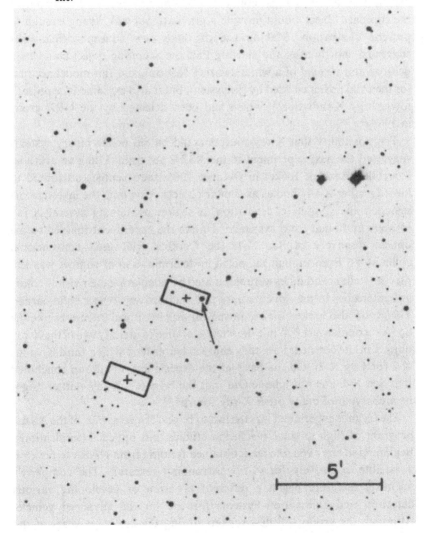

In theory, the prospects of the model as an explanation for the energy source of X-ray emitters appeared excellent. As mass was 'sucked' from the large partner, its kinetic energy would rise proportionally with the mass of the compact, highly evolved star and inversely with its radius. For sufficiently dense and small stars, this meant a huge amount of energy could be released. In contrast to the method of energy generation for main sequence stars like the Sun – namely nuclear burning of hydrogen, which produced about 7 MeV per nucleon – the accretion process could provide up to 100 MeV per nucleon if matter fell onto a white dwarf or hypothetical neutron star with a mass the same as the Sun.[45] The model could also explain how a hot, thin gas could remain confined near an X-ray source: the compact object would provide a gravitational sink strong enough to prevent dissipation. Making use of these new concepts, Shklovsky suggested shortly after the meeting that the accreting object could be a neutron star instead of a white dwarf.[46] Not only did this model account for the total power emitted by the system, but it also explained the object's low-energy X-radiation,[47] which had been detected by the NRL group in 1965.[48]

The possibility that X-ray sources could be old novae binary systems suggested the next experiment in the AS&E program. Using an attitude-controlled Aerobee rocket in October 1966, the scientists attempted to identify other X-ray bodies as similar objects. This time the investigators searched the Cygnus constellation, a region containing several X-ray emitters and binary star systems.[49] Unlike the experiment that led to the optical discovery of Sco X-1, the October trial used conventional collimators. Even though the position determination of sources was less precise, stellar candidates were eliminated by calculating their visible colors and intensities in the same manner that had proven so successful earlier. The group also looked for old novae varying in optical intensities like the Sco X-1 counterpart.[50] While the scientists failed to identify two of the X-ray objects in the constellation, they enjoyed better luck with a third, Cygnus X-2 (or Cyg X-2). On the basis of the experiment, the group concluded that Sco X-1 was not unique and that the brightest X-ray emitter might be archetypal of many other X-ray sources.[51]

The Cygnus experiment was the last to be performed as part of the AS&E program of high angular resolution studies and optical identifications. Beginning in late 1966, the team obtained funding from NASA to develop a satellite exclusively for X-ray astronomy research. The company's scientists therefore began a different program of developing various detection and collimation systems for use on the advanced vehicle. Although the group still flew rockets during the last three years of the decade, it employed them largely to test these systems rather than to pursue new discoveries and identifications. Those would come later when the satellite had been launched into orbit.

As a result largely of the NRL surveys and the AS&E program of examining features of sources, X-ray astronomy had undergone a significant qualitative change by 1967. In terms of instrumentation, the groups had progressed from small and uncollimated Geiger counters to large proportional counters having remarkable angular resolution. Observational knowledge in the field had made important advances too. Scientists found about 30 X-ray sources, which were classified into at least three categories: supernova remnants; old novae; and extragalactic radio sources. The identification of some X-ray sources as novae held the most promise for resolving the field's central problem because it stimulated a spurt of theoretical investigations concerning the binary accretion model,[52] a model that looked especially promising because it could explain how X-ray objects released so much energy from dilute plasmas. In theory at least, the model looked desirable. It suggested to a *Time* magazine reporter that X-ray astronomy was 'coming of age.'[53]

Despite its attractiveness, the binary accretion model could not establish a solid foundation in the 1960s. Although theoreticians may have accepted it, experimentalists found it pressed with many difficulties. Optical observations of the counterparts of Sco X-1 and Cyg X-2 suggested, for example, that they might be eclipsing binary systems, but neither provided the unambiguous evidence needed for proof. In particular, neither system exhibited the periodic velocity and intensity variations expected by orbital motion of two stars passing in front of one another.[54] Failure to detect these effects could have resulted, of course, from the plane of rotation being perpendicular to the observers' line of sight. Still, because they lacked firm evidence of these variations, scientists could not establish the binary nature of X-ray sources.

9

More problems, new lines of research

Unlike the AS&E team, the NRL group did not develop instruments and techniques for use on a projected satellite. Still, it also departed from its original research program of surveying the sky. The diminishing prospect of discovering new sources using scanning techniques helped bring about this change. Using these methods, the group detected weak sources that shone about 100 times less intensely than Sco X-1. To observe dimmer emitters, scientists required larger counters and longer observation times. On a spinning Aerobee rocket, however, both factors of detector size and time had been maximized. By using attitude-controlled rockets to point sensitive instruments at small regions of the sky, they could (and did, as was the case with 3C273) reach another factor of ten in sensitivity for individual sources, resulting in the discovery of about 10 more objects by 1970. But to search this way for weaker sources would have been prohibitively time-consuming and expensive. It would have taken hundreds of finely collimated rocket-borne detectors over many years to examine the whole celestial sphere. Such studies were best left for satellite observatories.

In place of surveying programs, then, the NRL team – and some newer X-ray experimental groups – pursued different research routes until satellites became available in the early 1970s. Two fields proved especially significant. The first – pulsar research – gave X-ray investigators a satisfactory model for the perplexing Crab Nebula emitter. Perhaps all other sources could be subsumed by similar models, and the central problem of the field could be quickly resolved. Meanwhile, research on the diffuse background X-radiation led to the development of two explanations of the phenomenon, one more fruitful than the other. Both theories dovetailed nicely with the big bang cosmological hypothesis, but one explanation led to the stunning conclusion that enough intergalactic mass exists to 'close' the universe. If substantiated, the evidence for this last theory would demonstrate again the value of X-ray studies to other realms of astronomy.

Pulsar research

The important line of research that led to model making concerned stars producing radio emissions at regular intervals. Discovered in 1967 by Cambridge University astronomer Anthony Hewish and his graduate student Jocelyn Bell, the first such body appeared during a search for rapid time variations of small radio sources.[1] Although the periodic phenomenon occurring every 1.33 seconds did not indicate that 'little green men' from a distant civilization were trying to communicate with us, it did imply something almost as remarkable. According to theoretician Thomas Gold of Cornell University, the pulses arrived from a rapidly rotating compact star called a 'pulsar.'

Gold's pulsar theory was elegant in detail and, more importantly, quickly substantiated. First of all, Gold showed in 1968 that a pulsar could not be just any commonly known dense object like a white dwarf. Instead, it must be a neutron star whose small size prevented losses of mass to centrifugal forces. The object still released some matter, Gold argued next, but in a different way. High-energy particles were emitted into strong, asymmetric magnetic fields, creating radiation in the pattern of a revolving lighthouse beacon[2] (Figure 18). Coming only from a small portion of the star, this radiation broadcast the observed flashes of radio waves. Finally, Gold predicted that because it ejected matter, the star would exhibit a 'slight, but steady slowing down of the observed repetitiveness.'[3] The discovery of this effect in 1969 provided compelling evidence for Gold's neutron star explanation of the pulsar.[4] More generally, it meant that a neutron star, proposed 30 years earlier from a theoretical standpoint, had found an observational basis.[5]

Within a year of the first pulsar discovery, ground-based optical and radio astronomers observed scores more. But none coincided with known X-ray sources. And because surveys of the sky with pointed rockets became impractical, investigators did not actively search for X-ray emitting pulsars. The situation changed in 1969, however, after the detection of a radio pulsar in the Crab Nebula – a well known X-ray emitter.[6] In the words of NRL's Talbot Chubb, this discovery in radio and optical wavelengths set off a 'classic research competition' to observe pulses in X-radiation too.[7] Chubb and his colleagues won the race with an experiment performed on 13 March 1969.[8] A few weeks later, the MIT rocket group conducted a similar investigation with the same outcome.[9]

Both groups (and many others subsequently[10]) found that the X-ray period of 33 milliseconds matched that observed in visible and radio wavelengths, confirming that astronomers working in three spectral regions all observed the same object. Moreover, they discovered that the X-ray wavelengths contained the most energy: the pulsar's X-ray power exceeded its optical power by a factor of 200 and its radio power by more than 20000.[11] But not all the X-radiation came from the central pulsar. Only about 5% stemmed from the spinning object, while the remainder originated from the entire nebula in a fashion that was not immediately clear.[12] This last observation corroborated the result obtained by the NRL group in its 1964 Crab Nebula occultation experiment. At that time the team searched for a point source, but found none.

An old astrophysical problem reemerged with the discovery of an X-ray pulsar in the Crab Nebula. As early as 1953, Josef Shklovsky argued that synchrotron radiation produced the Crab's continuous visible spectrum.[13] The subsequent observation of radio signals from the object could also be explained by this process. But detecting X-rays from the source was hard to understand. Although scientists knew that X-rays could be generated when relativistic electrons spiraled in magnetic fields, they also realized that the particles would lose their energy within a year of the supernova outburst – in the case at hand, more than 900 years earlier. After 1964, however, new evidence suggested that the nebula did indeed emit

Figure 18. A schematic diagram of a pulsar as a magnetised rotating neutron star in which the magnetic and rotation axes are misaligned. From Longair, *High Energy Astrophysics*, p. 222.

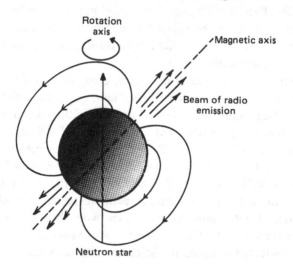

synchrotron X-rays.[14] How could this anomaly be resolved? What kind of unknown energy source could accelerate electrons for more than a year?

Gold's pulsar model provided the answers. As an ordinary star collapses to form a neutron star during a supernova explosion, its magnetic field collapses too, producing a much larger field strength – up to 10^{14} gauss for the Crab pulsar.[15] The extremely strong field constitutes the huge energy source which accelerates the electrons. Meanwhile, the charged particles come from the plasma ejected at the pulsar's poles. Consistent with the theory of the synchrotron mechanism, each electron will not radiate for more than one year. But because the pulsar always injects new electrons into the magnetic field, it continuously produces X-rays. The Crab's puzzling X-radiation had finally become understood.

But there was more good news, because the model also explained the nebula's diffuse X-radiation. As particles stream from the pulsar, they eventually enter a shock transition region, where they deposit their energy. This energy is then distributed to the rest of the nebula and becomes continuous (instead of pulsed) and extended radiation.[16] While some details about the interaction of the pulsar and nebula remained obscure,[17] the model nevertheless appealed to scientists. It received further substantiation from A. Finzi of NASA's Institute of Space Studies in New York City and Richard A. Wolf of Rice University, who calculated from the pulsar's increasing spin period that the object converted its rotational energy into just enough power to account for the nebula's expansion and all its electromagnetic radiation.[18]

The apparent success of the pulsar model in explaining the source, synchrotron spectrum, and spatial structure of the Crab's X-radiation suggested to some theoreticians that all X-ray objects might emit like the nebula.[19] Wallace Tucker of Rice University, for example, developed a basic model that consisted of a rotating neutron star losing mass in the presence of a large magnetic field. In the case of an extended X-ray source such as a supernova remnant, the energy of ejected plasma would be greater than the magnetic energy confining it, resulting in relativistic particles being able to leave the star's immediate neighborhood. In the case of a discrete star-like source, magnetic energy would dominate over the plasma's kinetic energy, and no mass would be lost from the magnetosphere. Although they would be confined to radiation belts around the star, the particles would still radiate and appear to be coming from an object having a small angular size.[20] Other theoreticians proposed pulsar models in which both synchrotron and bremsstrahlung radiation could be produced. These theories could subsume objects like Sco X-1 which emitted by the thermal bremsstrahlung process.[21]

Comprehensive in scope and successful in explaining the radiation emitted by the Crab Nebula, the pulsar model nevertheless could not advance much further in the late 1960s. Like the binary accretion hypothesis, the new model looked better in theory than in reality, because no other X-ray sources appeared to pulse like the Crab Nebula emitter. In searching for X-ray pulsars, the NRL group reanalyzed data from earlier observations. Still, it found no evidence of periodically variable emitters.[22] Radio and optical examinations of Sco X-1 in 1968 and 1969 also showed no periodic variations suggestive of a pulsar.[23] Further testing of the model appeared to require instruments having time resolution as short as one millisecond. Though these sophisticated detectors existed and could be ferried into space on sounding rockets, they could be used most profitably to survey the sky for X-ray pulsars if placed on large satellites (weighing from 4500 to 9000 kg each), as Friedman advocated in April 1970.[24] Such huge satellites, however, were not imminent in 1970 or even planned for X-ray astronomy research until the middle of the 1970s. Consequently, scientists shelved the pulsar theory.

Research on the diffuse X-radiation

Resolving X-ray astronomy's central problem involved more than just discovering the astrophysical processes that resulted in high-energy emissions from discrete sources. Although these objects were the subject of most studies, largely because they emitted more energy and could be observed more easily than the background, the diffuse X-radiation remained a major X-ray phenomenon requiring observational and theoretical treatment. But just as discrete sources appeared to need more than one type of explanation, so did the background radiation. By 1967, scientists realized that the diffuse 'hard' X-radiation (radiation having energies greater than a few keV) and the diffuse 'soft' X-radiation (radiation having energies as low as 1/8 keV) constituted different phenomena that perhaps had different origins in the universe.

Since the first experiment performed by the AS&E group in 1962, observations indicated the existence of a hard background flux in the energy region above 1 keV. Even as background 'noise' suppression techniques became more sophisticated and as fine angular resolution detector systems became widely employed, a real X-ray background always remained. The diffuse radiation continued to exist as observations extended into the range of hundreds of keV. Moreover, a single 'power law' spectrum described the background flux everywhere one looked, suggestive of a single, common origin.[25] Finally, this X-radiation appeared to be dispersed uniformly throughout the celestial sphere,[26] implying that it originated from discrete, remote sources outside the Galaxy. Similar to the hypothesis held by many radio astronomers attempting to understand the radio background,[27] this theory's simplicity attracted followers despite some experiments reducing its credibility. The problem of the diffuse hard radiation could, therefore, wait until individual sources were understood.[28]

Many scientists found the wait unnecessary as a more sophisticated, yet still relatively simple, model of the hard background radiation emerged. It postulated that low-energy intergalactic photons interacted through the inverse Compton effect with relativistic electrons emitted from remote sources, producing X-ray recoil photons. Proposed by James Felten and Philip Morrison[29] – the same theoreticians who suggested the effect as a mechanism for X-ray emission from discrete sources – the speculation appeared especially promising after 1964. In that year, Bell Laboratories' astronomers Arno Penzias and Robert Wilson discovered radio noise emanating from all parts of the universe.[30] Although the radiation's spectrum could not be measured precisely, it appeared similar to emission from a blackbody having a temperature of around 3 K. Most importantly, it supported the big bang cosmological theory, which suggested that all the matter in the universe was once contained in an intensely hot fireball. From the fireball's explosion 20 billion years ago, matter and radiation dispersed throughout space. Since then, the radiation cooled to 3 K, producing radio waves coming from all parts of the universe. Most importantly for the X-ray theoreticians, the big bang hypothesis implied a high density for the 3 K photons. This in turn made it likely that energetic electrons, emitted by radio galaxies, for example, would find enough photons with which to interact.[31] While some discrepancies between theory and observation remained, scientists averted them by resorting to cosmological models in which faraway radio galaxies emitted larger electron fluxes than those nearer the Earth. In general, the theory accounted nicely for the spectrum of high-energy diffuse X-radiation. Moreover, it united various astronomical phenomena into one coherent theory, making it appealing throughout the decade.

Success in explaining the hard background radiation suggested to many theoreticians that the low-energy background radiation could have a similar extragalactic origin. Despite its attractiveness, this straightforward speculation could not be easily tested because interstellar matter absorbs soft X-rays. Although the galactic disc, and hence all space, is essentially transparent to X-rays having energies greater than 4 keV, less energetic radiation is attenuated by the interstellar medium.[32] At energies less than 0.5 keV, observations can be made only within 100 parsec of the Earth toward the galactic plane, where neutral hydrogen and other elements block the radiation.[33] These facts meant that even if soft X-rays had an extragalactic origin, they would not be equally distributed throughout the celestial sphere. Instead, they would produce a nonsymmetrical 'latitude effect:' intense soft radiation at the poles, where little matter blocks radiations, and gradually less radiation as one looked toward the galactic plane. If such an effect could be observed, however, it would lend credence to the view that all diffuse radiation – hard and soft – had a distant origin.[34]

The possible detection of the latitude effect held great interest because the soft X-radiation could furnish important cosmological information. In the well-received big bang theory, it was uncertain whether the universe would expand forever or be 'closed' and eventually collapse upon itself. The determining factor consisted of whether enough matter existed in the universe to halt the expansion by gravity. To many cosmologists, a closed universe seemed philosophically attractive because it sidestepped awkward questions of the universe's beginnings and destiny. The answer to both queries, assuming enough matter existed in space, was simply that the universe originated from a singularity and would eventually return to it, perhaps in a cycle of big bangs and contractions.[35] Although pleasing to theoreticians, the model of a closed universe had accrued no observational evidence. Optical and dynamical studies of galaxies indicated much too little mass.

Nevertheless, some scientists thought that the 'missing mass' might exist as gas between galaxies. One such scientist, the Princeton theoretical astronomer George B. Field, had already begun seeking ways to measure the density of intergalactic matter. By analyzing radio observations of intergalactic space, Field found that neutral hydrogen, while plentiful, had a density too small to halt the universe's expansion. Next, he and his graduate student Richard C. Henry considered the abundance of ionized hydrogen, which could not be detected by radio means. If the density and temperature of the gas were high enough, the astronomers argued, then soft X-rays might be emitted and be observable from spacecraft.[36]

Exciting as such observations might be, instrumentation difficulties hindered them. The thickness of beryllium windows typically used to cover Geiger counters (five one-thousandths of an inch thick on the first nonsolar X-ray detectors) posed the first snag, because soft X-rays could not penetrate them. Thin organic films such as mylar transmitted soft X-rays efficiently, but problems remained. Very fragile – only about one ten-thousandths of an inch thick – the windows often ruptured when accelerated in rocket launches. Even if they did not, low-energy electrons could easily traverse the films and mimic X-ray photons by producing counts in detectors. Complicating matters further, the windows let gas diffuse out of the counters, which meant that fresh gas had to be continuously pumped into the instruments. This required pressure regulators, filters, and extra 'plumbing' devices to be carried in rocket payloads.[37]

Although difficult and chancy to operate, the NRL group had used these gas flow counters several times since the late 1950s for solar X-ray observations.[38] Partly because it had experience with them, the team made the first soft X-ray surveys of the sky in June 1964 and April 1965.[39] In these experiments, though, they aimed to examine discrete sources rather than background radiation.[40] But new groups, such as those formed at the University of California at Berkeley and the University of Wisconsin, made soft X-ray measurements of the background radiation a major part of their activities. By specializing in this relatively unexplored area of X-ray astronomy, they hoped to avoid direct competition with established groups.[41]

Led by C. Stuart Bowyer, formerly a graduate student working with the NRL rocket astronomy team, the Berkeley group performed its first soft X-ray background study in December 1966. On charting X-ray counts on a map of galactic coordinates, the group discovered what appeared to be a latitude effect for soft X-ray attenuation, suggesting the extragalactic origin of the radiation.[42] Providing more evidence for its remote nature, the soft flux exhibited roughly the same spectrum as harder (and presumed extragalactic) diffuse X-rays.[43] Having joined the Berkeley Astronomy Department in 1965, Field examined Bowyer's data from the 1966 soft X-ray survey.[44] Unfortunately, these early measurements could not provide an unambiguous determination of the gas density.[45] The missing mass hypothesis remained unsubstantiated.

Even before the Berkeley scientists published their results, NRL's Friedman took an interest in the possible detection of a dense intergalactic plasma. To pursue this research in 1967, he hired Richard Henry,[46] who began work on a rocket study of the soft X-rays emitted by several discrete sources and the galactic pole region. Performed in early September 1967, the NRL experiment detected an unexpectedly large soft background flux from the polar region and a few discrete sources near the Galaxy's plane.[47] The group reasoned that the point emissions must have been from within the Galaxy, because otherwise they would not have been seen through the interstellar medium.[48] On the other hand, the polar X-rays could have come from beyond the Galaxy.[49] Assuming this, Henry then analyzed the X-ray spectrum's shape and determined that the temperature of the ionized gas, if it existed, would lie between 200 000 K and 1 000 000 K. One more calculation led to an estimate of about 5×10^{-6} protons cm^{-3} for the gas' density.[50] If these assumptions and calculations were correct, Friedman wrote in 1969, then 'X-ray astronomy has provided the first evidence of nearly one hundred times as much mass in the form of intergalactic gas as in the galaxies themselves' – just enough mass, in fact, to close the universe.[51] This extraordinary, though still tentative conclusion, held enough significance to be reported almost as fact by *Time* magazine. A picture of a proud Richard Henry accompanied the article's title: 'Mystery of the Missing Mass.'[52]

While an exciting result, the 'discovery' of the missing mass rested on the fundamental assumption that soft X-rays had an extragalactic origin. The assumption received criticism from the X-ray astronomy group at the University of Wisconsin. Led by William Kraushaar, the group performed mapping experiments of the soft X-ray sky in September 1968 and December 1969. Though the observations provided good qualitative evidence for the latitude effect, they also suggested a much higher soft X-ray intensity in the galactic plane than expected if the radiation had come from beyond the Galaxy.[53] Other groups investigating the soft X-ray background arrived at similar results, but they accounted for the planar radiation by assuming a nonuniform distribution of interstellar gas and the existence of unresolved sources.[54] The Wisconsin scientists, however, remained skeptical. They argued that the general latitude effect could not be reconciled with quantitative models and hence, the assumption of the extragalactic origin of the soft X-radiation should not have been accepted so readily.[55]

In an attempt to end the debate concerning the soft background's origin, the Wisconsin group designed a sophisticated experiment for scanning the sky near the Small Magellanic Cloud.[56] The nearest extragalactic body, the Cloud was sufficiently large to be resolved clearly by detector systems and was relatively unconcealed by neutral hydrogen.[57] If most of the soft radiation originated from an extragalactic plasma, the scientists reasoned, then they would observe diffuse radiation obscured by the Cloud. Performed in June 1970, the experiment resulted in a null observation; neither a shadow, which would have conclusively demonstrated the extragalactic nature of the soft radiation, nor an emission appeared above the background level.[58] Although possible that the Cloud emitted soft X-rays of the exact intensity and spatial distribution as required to fill in the expected absorption of extragalactic radiation, this supposition was *a priori* unlikely.[59] The simplest interpretation was that most, if not all, of the soft diffuse X-rays had a local, galactic origin.[60]

Most X-ray astronomers, however, discounted this conclusion. The extragalactic hypothesis remained popular because it accounted for the large-scale observations of the latitude effect, and it encompassed numerous attractive cosmological principles.[61] Additionally, models could be invented to include sufficient clumping of interstellar gas to save the hypothesis.[62] Despite some dissent from members of the Wisconsin group, then, the soft diffuse X-ray flux was considered in late 1970, and for many years after, to have a remote and cosmological origin, just like the hard radiation.

The state of the art in 1970

After eight years of experimental work and various research programs attacking the field's central problem, X-ray astronomy remained in a state of flux. As for understanding how discrete sources converted energy into radiation – one aspect of the problem – scientists enjoyed some success with processes that led to thermal bremsstrahlung and synchrotron radiation. Relative success in determining how sources dissipated energy, however, was not the same as discovering how the objects had so much energy to release in the first place. A theoretical understanding of this aspect of the central problem proved considerably more difficult. The stationary neutron star model, although attractive when scientists had little observational information, was nevertheless quickly disproven by the Crab Nebula occultation experiment of 1964. And while the binary accretion model became popular among theoreticians, experimentalists discredited it for lack of substantiation.

Even if consensus was absent elsewhere, the 1960s produced two qualified success stories, the first coming soon after the discovery of pulsars in 1967. The model for a rotating magnetized star, elucidated initially to account for observations made by radio astronomers, also satisfactorily explained many details about the Crab Nebula's X-ray generation and emission processes. But evidence for the pulsar model did not materialize for other X-ray objects. The second accomplishment consisted of determining the extragalactic origin of the high-energy background X-radiation. As with pulsars, however, the success would not be extended without debate to other phenomena such as the low-energy background radiation.

X-ray astronomers found no solutions to the central problem in the 1960s partly because of the state of technology in the field. While impressive progress had been made in observing celestial X-rays since 1962 – by using more sensitive detectors, small-angle collimators, and attitude control systems – experimental developments necessary for advancing theoretical understanding were limited by the capabilities of sounding rockets. No matter how well stabilized, the vehicles could obtain information from sources for no more than just five minutes, rendering some measurements impossible.

Good spectral data, for example, proved hard to obtain in rocket experiments. The main problem consisted of acquiring enough photons to assemble a spectrum in a wide energy band. Because the sources had such low intensities when compared to the solar X-ray flux, direct observations of hours rather than minutes were necessary. Of course, scientists obtained some low-and high-energy spectra from sources, but their interpretation remained ambiguous. As noted by Lodwyk Woltjer, a Columbia University theoretical astrophysicist, almost any spectrum from a discrete source could be fitted to a model of bremsstrahlung or synchrotron emission with the proper choice of a few key parameters.[63] While convincing theoretical proof of a thermal mechanism (either blackbody or bremsstrahlung) would come from observing spectral lines from highly ionized atoms,[64] these measurements in practice were difficult to make. The positive identification of lines required good energy resolution of proportional counters, a feature lacking in most detectors because, again, few photons availed themselves for analysis. Several groups' attempts to observe iron lines in the spectrum of the brightest source, Sco X-1, tended to be inconclusive.[65] Line measurements of the soft diffuse background would also have had great cosmological implications, by providing evidence that the intergalactic plasma radiated by thermal bremsstrahlung. Nevertheless, no such observations had yet been made because of the radiation's low intensity. Additionally, if experimentalists had found a source's X-ray emission to be polarized, they could have conclusively identified it as synchrotron radiation.[66] Practical difficulties resulting from the use of short-lived rockets made such observations difficult too. Columbia University scientists attempted to measure polarized radiation from Sco X-1 in 1968, but they obtained no conclusive results.[67]

The near complete reliance on rockets also limited the number of optical identifications possible. As seen in the cases of the Crab Nebula and Sco X-1, observations of X-ray emitters in different spectral regions provided new avenues for approaching the field's central question. While the modulation collimator technique used by the AS&E and MIT groups worked for locating Sco X-1, scientists employed it only once again.[68] Because each wire grid in the collimators cut in half the effective area of the counters, only strong sources could be studied with the instrument.[69] Even by using advanced background suppression techniques, rocket-borne detectors could not observe X-ray emitters long enough to make the modulation collimator method practicable in any other cases.

Finally, the prospect of discovering new X-ray sources from rockets appeared unlikely. As seen, Friedman terminated the NRL surveying program, whose main purpose was to locate previously unobserved X-ray emitters, because he realized that even the most sophisticated detectors carried on rockets could not reveal the existence of many more sources. And his judgement proved valid. From 1962 to 1967 the pace of discovery had been swift, and about 30 sources were discovered, most by NRL investigators. In the next three years less than a dozen new emitters were observed by *all* experimental X-ray groups. During this period of diminishing returns, Friedman noted in 1969, 'the pace of discovery has ... slowed down considerably, and the next great surge in detection of new sources will have to await the use of satellites.'[70]

The potential usefulness of satellites for X-ray astronomy observations had been considered well before 1969. As early as 1960, Philip Fisher of the Lockheed company suggested placing detectors on artificial moons. And in 1966, the Lockheed group actually received the first opportunity to perform a satellite-borne experiment. As noted in Chapter 4, NASA supported Lockheed programs for devising X-ray detectors and spectrometers. In 1965, when an unexpected vacancy arose on the first Orbiting Astronomical Observatory – the heaviest (around 1800 kg) and most sophisticated scientific satellite yet – the space agency invited the Lockheed group to provide some proportional counters having effective window areas of 700 cm^2.[71] Although the team had the 'chance to scoop the world' in X-ray astronomy by being the first to make sustained observations of sources with a highly sensitive detector, it encountered misfortune.[72] Launched in April 1966, the spacecraft suffered electrical malfunctions immediately upon achieving orbit, causing the failure of all experiments on board.[73] Throughout the late 1960s, the Lockheed scientists experienced continued bad luck in flying other rocket payloads, resulting in little observational data ever being acquired. This explains why the group received scant mention in previous chapters.

Much more fortunate in the long run, the AS&E group launched its X-ray astronomy satellite in 1970 and obtained data from hundreds of new sources, many which revealed hitherto unnoticed features. These discoveries finally enabled scientists to acquire a satisfactory solution to the field's central problem as it applied to discrete sources.

10

Uhuru

From a modified oil rig off the sandy coast of Kenya, engineers launched the first X-ray astronomy satellite into orbit around the equator (Figure 19). Nicknamed 'Uhuru' – meaning 'freedom' in Swahili – in honor of its launch on Kenyan Independence Day, 12 December 1970, the observatory passed under the dangerous radiation belts surrounding the Earth and detected 339 discrete sources in its 2 1/4 year lifespan. Although impressive, the acquisition of a huge list of emitters was not the satellite's most significant accomplishment. Foremost was its discovery of previously unobserved variations in the intensities of just a few objects. Indicating unambiguously that some sources constituted double star systems, the variations finally provided evidence in support of the binary accretion model. Thereafter, in what one theoretical astrophysicist called a 'bandwagon' effect, scientists interpreted almost all galactic X-ray emitters with the binary model.[1] This consensus over a conceptual framework signified that a major part of X-ray astronomy's central problem had been resolved.

Instruments and observations

Conceived of by Riccardo Giacconi in 1963, the first X-ray astronomy satellite had the mission of producing a catalog of new sources and of observing their positions, intensities, and spectra. When Giacconi proposed the project to NASA, only two objects – Sco X-1 and the Crab Nebula – had been detected. By studying more X-ray emitters and some of their features, the AS&E scientist hoped to understand how the objects generated and emitted so much energy.[2] Funding for the program began in late 1966, soon after the failure of the ambitious first Orbiting Astronomical Observatory. The disaster emphasized the potential consequences of cutting space exploration costs by housing several experiments on one large, untested vehicle.[3] Contrasting this approach, the AS&E

satellite carried only one experiment for one research objective. As 'SAS-1,' NASA's first 'Small Astronomy Satellite,' Uhuru became the prototype for subsequent single-function orbiting observatories.[4]

The Uhuru satellite carried advanced, though still conventional, instrumentation. Not much heavier than a typical sounding rocket payload, weighing only 64 kg, the detection devices included a pair of proportional

Figure 19. Launch of SAS-1, 'Uhuru,' from the San Marco platform off the coast of Kenya on 12 December 1970. Courtesy of the National Aeronautics and Space Administration.

counters having a surface area of about 840 cm² each. Mounted back to back, the counters could be used in an anticoincident mode to reduce background noise created by charged particles.[5] To increase sensitivity even more, the AS&E scientists used a novel background rejection technique called 'pulse shape discrimination.' Modified for use in nonsolar X-ray astronomy by Paul Gorenstein and tested in rocket experiments in 1967, the system distinguished narrow X-ray pulses from wider (or longer lasting) pulses produced by gamma rays or electrons (Figure 20). By electronically shunting events having wide pulses, it eliminated up to 90% of the non-X-ray background.[6] Collimators having views of 1/2° by 5° and 5° by 5° completed the detection system (Figure 21). The different fields of view gave fine angular resolution to one counter and high sensitivity to the other.[7] Launched on rockets before 1970, the instruments operated properly and gave the AS&E scientists confidence in the entire system.

When carried in a satellite, the instruments made three types of measurements with greater precision than those flown on rockets. First, they detected weaker sources. Even though the vehicle spun, its instruments could repeat observations of objects and accumulate photon counts from them on each scan. By doing this, Uhuru discovered sources having intensities 10 times fainter (as low as 10^{-11} erg cm^{-2} s^{-1}) than those observed with the best instruments carried on stabilized rockets. This greater sensitivity enabled Uhuru to increase the number of known X-ray sources by almost an order of magnitude[8] (Figure 22). Secondly, by scanning frequently over the same object, the satellite's detectors made

Figure 20. Diagrams illustrating the principles of rise-time or pulse-shape discrimination between X-rays and cosmic rays. From Longair, *High Energy Astrophysics*, p. 89.

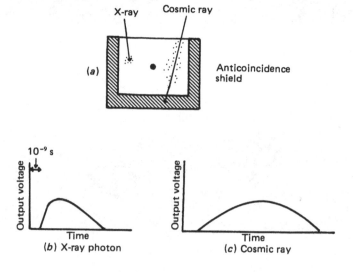

Figure 21. The Uhuru X-ray satellite showing (*a*) the lay-out of components and (*b*) the beam pattern of the telescope on the sky. From Longair, *High Energy Astrophysics*, p. 91.

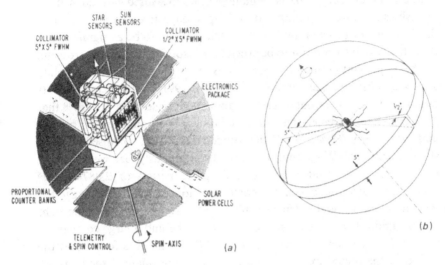

Figure 22. The 339 sources detected by Uhuru are displayed in this sky map of galactic coordinates. The size of the symbols representing the sources is proportional to the logarithm of the peak intensity. Reprinted courtesy of Riccardo Giacconi and the *Astrophysical Journal Supplement Series*, Vol. 38, p. 409, published by the University of Chicago Press. © 1978 The American Astronomical Society.

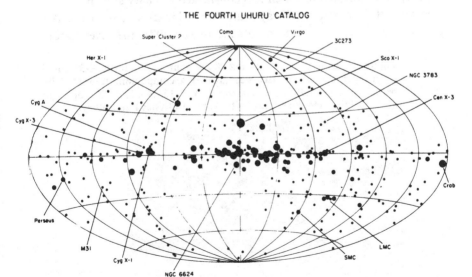

better position measurements. Of the more than 300 sources observed, location determinations of about 100 were accurate enough to suggest visible and radio counterparts.[9] Finally, Uhuru measured variations in the intensities of sources for long periods – up to days at a time. Even though this last type of observation had not been planned to be as significant as the other two, it turned out to be crucial for resolving the central problem.

The important series of observations began soon after the satellite achieved orbit. Between January and May 1971, Uhuru's instruments examined the source Centaurus X-3, an emitter discovered in 1967[10] that now exhibited extremely regular variabilities. With an intensity pulsing uniformly every 4.8 seconds,[11] the source resembled a collapsed rotating star like a pulsar.[12] While being monitored continuously for a week in May, the source next revealed a secondary pulse period of 2.087 days.[13] A common astronomical characteristic, this longer periodicity suggested that investigators had detected a binary star system in which the light output diminished when a small object passed behind a large companion (Figure 23). In support of this view, the AS&E scientists found that the period of the higher-frequency pulses varied sinusoidally, exactly as expected from a system in which orbital motion was occurring.[14] Ground-based astronomers provided even more confidence in the interpretation when they discovered that the optical counterpart of Cen X-3, a massive blue supergiant star, exhibited the same pulse periods and variations.[15] Thus, the X-ray observations were not instrumental effects from Uhuru. Rather, they indicated that Cen X-3 constituted a binary system.

Positive that at least one X-ray source was a binary system, the AS&E group used Uhuru to search for others. In late 1971, the investigators

Figure 23. Eclipse of the pulsing X-ray source Centaurus X-3 caused by the orbit of a compact star around a companion. Reprinted by permission from *APL Technical Digest*, Vol. 15, No. 1, p. 22. Copyright © 1976 Johns Hopkins University Applied Physics Laboratory.

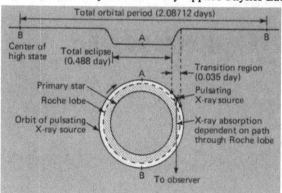

discovered Hercules X-1 (or Her X-1), an object that had periodicities of 1.24 seconds and 1.7 days.[16] Identified in optical wavelengths in mid-1972, the counterpart was a blue star having a mass about double that of the Sun. The scientists substantiated the identification by observing the pulse periods in optical wavelengths, which matched the X-ray periods exactly.[17]

The binary character of Cen X-3 and Her X-1 had been relatively simple to ascertain from their periodic emissions. It was more difficult to fathom Cygnus X-1 (or Cyg X-1), first observed by the satellite in December 1970, whose intensity fluctuated rapidly on as short a time scale as could be recorded with the Uhuru instruments (96 milliseconds).[18] Although not periodic, these rapid variations implied that the AS&E scientists had again detected a compact object. Shortly after these initial observations, the MIT rocket group studied the source with a proportional counter that resolved temporal events as short as one millisecond.[19] The same rocket experiment also located the object within a small field of view, leading ground-based astronomers to detect signals from a source soon identified as a massive blue supergiant star.[20] When observed in a spectroscope, the star exhibited Doppler shifts of its visible spectral lines, indicating that the emitter was indeed a binary system.[21] No long-period X-ray pulses had been observed simply because the system was inclined obliquely to the observer's line of sight.

The binary accretion model

In developing a model for the basic energy source for the X-ray emitters, theorists drew from earlier observational and theoretical work. Because both Cen X-3 and Her X-1 exhibited regular pulsations, theorists first considered the mechanism present in the Crab Nebula, namely that the electromagnetic radiation resulted from the conversion of the pulsar's rotational energy. If the same physical process occurred in the newly discovered pulsing sources, however, the periods would exhibit noticeable increases within a few months to account for their large luminosities of about 10^{37} erg s^{-1}.[22] Accurate measurements indicated an opposite effect: instead of increasing, the periods decreased slightly.[23] Scientists therefore discounted the hypothesis.

While the pulsar theory could not explain the copious X-ray production in binary systems, the accretion theory appeared more plausible. Proposed as an energy generation mechanism in 1966 following Sco X-1's identification as a possible double star system, the accretion model received new attention after Uhuru discovered binary X-ray sources. In a set of seminal papers written in 1972, English, American, and Russian theoreticians considered two cases in which a highly evolved compact object orbited a more massive companion.[24] In the first case, which subsumed the Cen X-3 and Her X-1 systems, the small and dense body was a neutron star, which accreted matter from the stellar wind or outer atmosphere of its companion. Matter would not just fall onto the neutron star, however. Because the infalling mass would carry angular momentum, it would form an 'accretion disc' around the neutron star (Figure 24). It is to this disc that the infalling matter deposits its gravitational potential energy, which would be converted into high temperatures and X-rays through the thermal bremsstrahlung process. Such a spectrum was already noticed from the two objects. As angular momentum of particles in the disc dissipates through viscous interactions with the surrounding gas, material would spiral inward onto the neutron star. Channeled by the asymmetric magnetic field lines of the dense object, the plasma would eventually reach the surface at the magnetic poles, where two hot X-ray emitting spots would form. Rotation of the object would cause the rapidly pulsed emissions.[25] Finally, this binary accretion theory predicted that because the energy source of the X-rays is gravitational potential energy – not rotational energy – the neutron star would not slow down like the Crab Nebula pulsar. In fact, since the neutron star would acquire matter having angular momentum, it would accelerate slightly, an effect already noted for the dense object in the Cen X-3 system.[26]

Figure 24. Accretion of matter onto a neutron star. On the left, the atmosphere of the large star overflows its critical surface and transfers mass to the neutron star. On the right, the stellar wind of the large star provides the matter that is accreted by the compact body. Reprinted by permission from *APL Technical Digest*, Vol. 15, No. 1, p. 20. Copyright © 1976 Johns Hopkins University Applied Physics Laboratory.

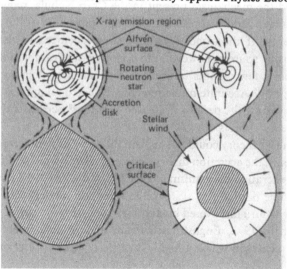

In the second case of Cygnus X-1, theoreticians interpreted the compact body as a black hole, a star ending its evolution with a density so great that even light cannot escape its surface.[27] Like a neutron star, a black hole is created during a supernova explosion, but only when the contracting core of the original star has a mass exceeding the theoretical maximum for a neutron star (a few solar masses). Because the compact body in the Cygnus X-1 system had a mass of greater than five solar masses, a fact learned from spectroscopic observations in optical wavelengths, it made a likely candidate for the first black hole ever observed. Even so, the object would still be accompanied by an accretion disc. At the disc's inner edges, infalling gas would become hot enough to generate X-rays. But because the black hole is more massive than a neutron star, it would need only a modest accretion rate to produce an X-ray luminosity of around 10^{37} erg s^{-1} – the power emitted by the Cygnus source. The accretion would also account for the source's observed thermal bremsstrahlung spectrum.[28] Other than the masses of the collapsed bodies, the systems differed in their pulse periods. Because a black hole could not have any asymmetrical features, such as magnetic field lines unaligned with its rotational axis, it could not pulse regularly. Instead, the turbulent fluid disc around the body would probably assure a rapid, but non-periodic, variability in intensity.[29]

In explaining the variabilities, luminosities, and spectra of these three X-ray sources, the binary accretion model appeared successful, yet simple. Confident of the model's ability to account for X-ray features exhibited by other sources, Princeton theoreticians Remo Ruffini and Robert W. Leach suggested a general classification scheme in 1972. In it, they identified all pulsing X-ray sources as binary systems containing neutron stars and those having irregular pulsations as black hole configurations.[30] Data obtained from Uhuru, succeeding satellites, and rockets seemed to verify this scheme as scientists discovered more binary systems and as mass determinations for the compact objects became more precise.[31]

These explanations of X-ray sources gained further credence after Dutch astronomers Edward P. J. van den Heuvel and John Heise proposed an evolutionary scheme in 1972 that considered compact stars in binary systems[32] (Figure 25). In the beginning, a typical configuration consisted of a closely bound pair of 'normal' stars with 16 and 3 solar masses. Evolving independently, the more massive primary star would exhaust its hydrogen supply in about seven million years, at which time it would begin expanding. During the next 20 000 years, the larger star would transfer 12 solar masses to the secondary. Much less massive now (only 4 solar masses), the primary would explode as a supernova in less than two million years, creating in the process a residual neutron star (or a black hole, if the remaining mass were large enough). Fifteen million years into its life, the other star would end its hydrogen burning phase and become a blue supergiant, which would then lose mass to the compact object as its atmosphere expanded or by emitting stellar wind. During this phase, lasting only between 20 000 and 50 000 years, the system would emit X-rays with a strength of 10 000 stellar luminosities. The copious production would end when the supergiant grew larger and exchanged more of its mass to the compact object, at which time the high density of accreted mass near the small star would absorb the X-rays and prevent them from leaving the neighborhood.[33]

Figure 25. Suggested evolutionary history of a system like Centaurus X-3. The primary star is on the left, and the vertical line indicates the center of mass. Reprinted by permission from *Nature Physical Science*, Vol. 239, No. 92, p. 68. Copyright © 1972 Macmillan Journals Limited.

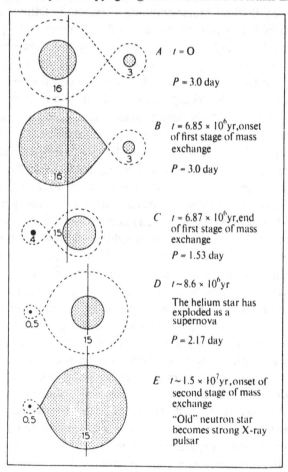

A $t = 0$

$P = 3.0$ day

B $t = 6.85 \times 10^6$yr, onset of first stage of mass exchange

$P = 3.0$ day

C $t = 6.87 \times 10^6$yr, end of first stage of mass exchange

$P = 1.53$ day

D $t \sim 8.6 \times 10^6$yr

The helium star has exploded as a supernova

$P = 2.17$ day

E $t \sim 1.5 \times 10^7$yr, onset of second stage of mass exchange

"Old" neutron star becomes strong X-ray pulsar

This evolutionary picture of X-ray binary systems, though revised since 1972 in the light of further developments in the field, accounted well for a number of phenomena associated with X-ray sources. First, the scheme predicted that compact objects in binary systems were comparatively old, having originated in supernova explosions millions of years before the X-ray production stage. It therefore explained the long pulse period of Cen X-3 – much longer than that known for any pulsar like the one in the Crab Nebula. The old age of the compact object also accounted for the absence of lingering gas shells around the system; they would have dissipated into space long before X-ray emission began. Next, the picture explained the association of massive blue supergiants with two of the three binary systems – Cen X-3 and Cyg X-1.[34] Finally, because the X-rays were supposedly produced during just a small fraction of the system's total lifetime, only a few close supergiant binary systems would exist as X-ray sources at any time. Since massive close binaries are rare in the Galaxy in the first place, this explained the sparsity of observed sources.[35]

With the development of such evolutionary schemes, the binary accretion model became an accepted solution to the field's central problem as it pertained to discrete galactic X-ray objects (but not to extragalactic sources such as quasars). Simply put, the model explained neatly both the source of the tremendous energy contained in X-ray systems and the mechanism for its conversion into radiation. Subsequent observations and theoretical studies of other binary systems lent more strength to it. Several interpretive problems still existed, of course, such as the origin of the soft background X-radiation and even the source of X-rays from Sco X-1. Still, the theory provided a basis for understanding many newly observed phenomena. The discovery in 1975 of objects whose X-ray intensities peaked in recurring patterns from their normal steady level was one example. Almost immediately after their first detection, scientists satisfactorily described these X-ray 'bursters' as accretion phenomena in binary systems.[36]

The widespread acceptance of the binary accretion model in the early 1970s marked the end of a period of searching for the central question's answer. The direct impetus for finishing this phase in X-ray astronomy's development was the use of a satellite as a new technological research tool. Like other nascent fields of science, such as high-energy particle physics in the first four decades of the twentieth century, X-ray astronomy required a store of fundamental experimental data before a solid conceptual framework could be established. In the mere two or three hours of observation time provided cumulatively by rockets during the 1960s, scientists had not acquired some of these basic facts, most notably those concerning the binary nature of a few X-ray sources. By delivering these facts in a conclusive manner, Uhuru did what no number of further rocket experiments could have duplicated. This does not imply that scientists foresaw the central problem's resolution as soon as Uhuru achieved orbit. After all, they did not realize exactly what crucial measurements were necessary before the satellite began observing sources. Nevertheless, X-ray investigators knew from their experience with modulation collimators, attitude-controlled rockets, and other devises, that as their tools gained sophistication, so did their conceptual understanding. As a dramatic advance in research technology, Uhuru should have been expected to produce important results as well. Indeed, several scientists anticipated making important observations of some kind with satellites, as is evidenced by statements published before 1970.[37]

Figure 26. An X-ray photograph of the Crab Nebula, taken through the 0.6-meter diameter telescope on board the Einstein Observatory (HEAO-2). The bright object in the center is the Crab pulsar. The photograph reveals a complex transfer of energy from the pulsar to the rest of the supernova remnant that is still poorly understood. Courtesy of the National Aeronautics and Space Administration.

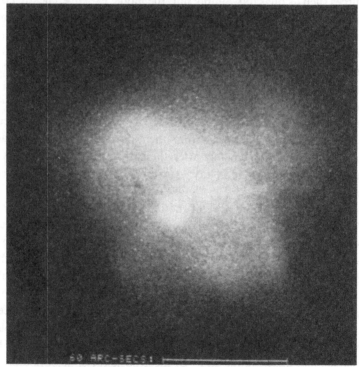

While signaling the end of one phase of X-ray astronomy's history, Uhuru ushered in another. During this phase, lasting until 1978, satellite technology predominated in observing programs of X-ray phenomena. In these eight years, seven new vehicles carrying X-ray detectors orbited the Earth; for several months in 1975 and 1976, six satellites operated simultaneously.[38] Satellite-borne instruments meant that scientists no longer had to settle for brief glimpses of the X-ray universe. They could now 'stare' at phenomena just like their ground-based colleagues. As expected, the new activities quickly yielded a wealth of data on discrete galactic and extragalactic sources as well as on diffuse X-rays emitted from clusters of galaxies. Together, the observations and interpretations contributed to the conception, derived first by radio astronomers and then by the early X-ray investigators, that the cosmos is not as calm and simple as it looks to optical observers. Instead, the invisible universe contains events of astonishing violence and high energy. No longer did astronomers consider explosive phenomena, such as those occurring in quasars, pulsars, and binary systems that included compact bodies, as inconsequential anomalies in a normally serene cosmos. Rather, they constituted important events that were crucial to the understanding of the energetics, dynamics, and evolution of the entire universe.

But the violent, highly energetic universe no longer remains the only subject for study in X-ray astronomy. Just as Uhuru stimulated a qualitative change in the field's understanding and subsequent research, so has the second High Energy Astronomy Observatory (HEAO-2), nicknamed 'Einstein.' The major new tool carried by the vehicle was the imaging 0.6-meter diameter X-ray telescope conceptualized by Giacconi in 1959 as a powerful X-ray 'funnel.' Launched in November 1978, the satellite inaugurated X-ray astronomy's third phase, characterized by studies of celestial objects that interested traditional astronomers. Besides taking pictures of explosive extragalactic objects and supernova remnants (Figure 26), the instrument examined calmer phenomena such as the coronal X-ray emission from Sun-like stars, the formation of new stars from gas clouds, and even the structure of magnetic fields around Jupiter and Saturn.[39] In other words, X-ray researchers observed almost the entire gamut of astronomical events. According to Giacconi, X-ray investigations have become as important to the understanding of events in the universe as those made in optical and radio astronomy. Because of this, X-ray astronomy today is one leg of 'the three-legged monster of modern astronomy.'[40]

SECTION IV

EPILOGUE

11

Success and frustration

In 1950, the prominent radio astronomer, Martin Ryle, commented on some initial rocket experiments made to study celestial phenomena. Although he witnessed how observations made in a previously unexplored wavelength region expanded scientists' view of the universe, he was not encouraged about the future of high-altitude astronomical investigations:

> Whilst some important results are likely to be obtained by the development of further rocket-borne apparatus, it seems unlikely that such brief and expensive experiments will play a major part in the growth of observational astronomy.[1]

This book has been written in part to demonstrate that Ryle's assessment was incorrect. As seen in the case of X-ray astronomy, brief and expensive experiments carried out in the 1960s and early 1970s made significant contributions to many aspects of celestial research.

Ryle was not an unperceptive man, and he has not been quoted to belittle his substantial scientific abilities. However, he failed to foresee important reasons why some scientific fields, difficult and expensive as they might be, would evolve into major areas of investigation. Most importantly, he had no way of anticipating how scientific research in the United Sates would be used to satisfy political needs. But this was exactly what occurred in the cases of X-ray astronomy and other fields of space science. As a response to the perceived crisis precipitated by the Sputnik satellite launches in 1957, government planners and political leaders chose to pursue an expensive program of space exploration.

With the country's image as the world's mightiest scientific and techno-logical force at stake, politicians agreed to do almost anything necessary to foster dramatic scientific accomplishments in space. In X-ray astronomy, the way toward discovering exciting phenomena consisted of giving scientists money to develop qualitatively different forms of research technology. During the 1960s, X-ray investigators explored the sky initially with simple Geiger counters. As these instruments revealed the first intense X-ray sources, scientists began asking questions about the physics of the objects. Consequently, they required and devised sophisticated instruments, such as large proportional counters, which they sometimes used in conjunction with modulation collimators. In many cases, investigators placed these improved devices on more versatile vehicles, such as attitude-controlled rockets and a satellite. As shown, the use of these advanced tools almost inevitably led to crucial discoveries and leaps forward in conceptual understanding. The use of a modulation collimator on a stabilized rocket in 1966, for example, enabled the precise location and identification in optical wavelengths of Sco X-1. And the employment of large-area proportional counters on the Uhuru satellite in the early 1970s revealed the binary nature of some X-ray sources.

The problem with this type of progress, dependent on the continuous sophistication of experimental technology, was that dramatic qualitative advances became increasingly expensive. In the first years of X-ray astronomy research, a typical sounding rocket and its experiment-bearing payload could be launched for about $125 000. To Ryle, this would have been an expensive experiment. Compared to the cost of five minutes of ground-based observation time, this was indeed a huge sum. The amount did not seem significant for space exploration, however, especially when compared to the cost of the manned lunar landing mission – more than $21 billion by 1969[2] – or even to later research in X-ray astronomy. The AS&E-MIT modulation collimator experiment of 1966, for example, cost more than $362 000.[3] Still more extravagant, Uhuru required about $37 million to build and place in orbit.[4] Though expensive, the development of these research tools was generously supported by the government from the 1960s until the early 1970s.

Of course, X-ray astronomy was never the subject of agitated public policy debates in the 1960s. It was simply too small a program within the multi-billion dollar space budget. While NASA officials at Congressional hearings would occassionally report on results in the field, few members of the House Committee on Science and Astronautics ever questioned the agency's funding for it. This comes as no surprise considering that in 1969 alone, NASA officials reported on 19 major research and development programs (comprising 82 different projects) in a period of no more than eight days.[5] X-ray astronomy, with expenditures of less than $3 million that year, was still a minor project in a $4 billion space program.[6] Members of Congress simply had better things to do with their limited time. Consequently, X-ray astronomy investigators benefited from the influx of new money for the space program but without much questioning from their benefactors. Funding decisions were left to NASA program administrators, who were assisted by in-house committees and outside groups such as the President's Science Advisory Committee and the National Academy of Sciences.

During the 1970s, however, X-ray astronomy's successes made it much more visible. With a large segment of the physics and astronomy communities supporting more ambitious projects in the field, NASA began the High Energy Astronomy Observatory program to follow up on Uhuru and other small-satellite programs. In lengthy testimony to Congress in 1971, agency officials argued that the program would be the most important new project of the year and that it deserved $13.6 million for fiscal year 1972.[7] Until 1977, when the second of four huge orbiting observatories was to be launched, the program would cost between $180 to $250 million. No longer an insignificant amount, the program's funding nevertheless received enthusiastic approval by Congressional oversight committees. The enhanced exposure and funding marked X-ray astronomy as a 'big' science.

Though support for the field had fortuitously coincided with major public policy initiatives in the past, X-ray astronomy's 'unveiling' in the early 1970s could not have been more poorly timed. In January 1973, the Nixon administration imposed budget cutbacks on government agencies, partly as a result of continued high spending for the Vietnam war. The space agency was an easy target for budget trimming. Losing political clout after completing its major mission of landing a man on the Moon in July 1969, NASA also suffered from the general public backlash against science and technology, which were often blamed in the Vietnam era for environmental pollution and social malaise.[8] Consequently, with NASA ordered to cut 1973 expenditures by $179 million,[9] administrators sought to halt programs still in their early phases to minimize losses of prior investments. HEAO met this requirement exactly. Eliminated totally, HEAO was restructured later in 1973 on a much less ambitious scale. Funding dropped from a proposed $70 million in fiscal year 1974 to just $5 million, with launch schedules pushed back about three years.

The remainder of the 1970s brought both success and frustration. On the positive side, American and European scientists launched nine new X-ray astronomy satellites and several suborbital vehicles between 1972 and 1979. Among these were the two redesigned High Energy Astronomical Observatories. By any standards applied, all the satellites performed spectacularly, and investigators eagerly planned for future opportunities.[10] But the experience of funding during the decade tempered their optimism.

Although several scientific committees recommended placing extremely large and powerful X-ray observatories – HEAO's successors – in orbit by the late 1970s, none became realities. Cost overruns and delays caused by NASA's Space Shuttle siphoned money from all space science programs, and practitioners do not expect to launch the next generation of X-ray observatories until the late 1980s.

In the 1970s, then, X-ray astronomy emerged as a major scientific activity making overt demands on public funds. Consequently, it became subject to Congressional and Presidential scrutiny. The new situation called for different approaches to secure research opportunities: they now needed to engage in the politics of open debate just like any other interest group. Unlike the 1960s, a period of rapid expansion and easy support, the next decade was a more difficult one in which investigators resolved the field's primary questions less quickly than they desired. In working through good times and bad, X-ray astronomers learned first-hand the highly political and often frustrating way that scientists conduct research in modern America.

APPENDIXES

Statistics on scientists writing seven or more papers in X-ray astronomy 1962–72

Number of papers	Number of scientists who wrote this many papers	Names of scientists
7	16	V. M. Blanco (O), J. A. M. Bleeker (F), R. Brucato (O), A. Cavaliere (T), R. Cruddace, R. Doxsey, P. J. Edwards (F), J. L. Elliot (O), W. Forman, J. R. Harries (F), F. R. Harnden, H. S. Hudson, R. M. Pelling, G. Setti (F), C. M. Wade (O), K. Yamashita (F)
8	13	L. W. Acton, R. D. Belian, S. Biswas (F), D. Brini (F), G. Buselli (F), M. C. Clancey (F), J. E. Felten (T), F. Fuligni (F), T. Kato (F), W. Mayer, A. J. Meyerott, A. S. Prakasarao (F), G. Riegler
9	13	G. Chodil, P. J. N. Davison (F), W. D. Evans, P. A. Feldman (T), G. J. Fishman, R. C. Haymes, U. B. Jayanthi (F), G. K. Miley (O), A. R. Sandage (O), J. P. Stoering, R. M. Thomas (F), M. Wada (F), B. G. Wilson (F)
10	10	L. L. E. Braes (O), P. C. Coleman, J. P. Conner, B. A. Cooke (F), J. F. Dolan, R. J. Gould (T), W. L. Kraushaar, W. Liller (O), R. E. Price, Y. Tanaka (F)
11	15	A. N. Bunner, G. A. Burginyon, G. S. Gokhale (F), V. S. Iyengar (F), J. Kristian (O), P. K. Kunte (F), J. E. Mack, O. P. Manley (T), D. McCammon, J. E. McClintock, K. G. McCracken (F), S. Miyamoto (F), M. J. Rees (T, F), B. B. Rossi, P. Vanden Bout
12	8	R. J. Francey (F), W. E. Kunkel, R. K. Manchanda (F), B. Margon, H. Mark, R. Rodrigues, H. W. Schnopper, S. Sofia (T)

Appendix (*cont.*)

Number of papers	Number of scientists who wrote this many papers	Names of scientists
13	2	P. C. Agrawal (F), D. A. Schwartz
14	6	J. R. P. Angel, P. C. Fisher, R. M. Hjellming (O), T. M. Palmieri, U. R. Rao (F), C. D. Swift
15	11	G. R. Burbidge (T), S. S. Holt, M. Lampton, D. E. Mook (O), S. Murray, R. Novick, Y. Ogawara (F), J. S. Shklovsky (T, F), J. I. Silk (T), W. B. Smith, W. H. Tucker (T)
16	3	G. Garmire, R. C. Henry (T), J. R. Waters
18	5	G. G. Fritz, J. F. Meekins, K. A. Pounds (F), E. Schreier, B. V. Sreekantan (F)
19	2	R. J. Grader, S. Rappaport
20	4	G. W. Clark, R. W. Hill, M. Matsuoka (T, F), P. J. Serlemitsos
21	1	L. E. Peterson
22	3	E. A. Boldt, H. Bradt, W. A. Hiltner (O)
23	2	H. M. Johnson (O), W. H. G. Lewin
24	1	S. Hayakawa (T, F)
26	1	P. Gorenstein
28	1	M. Oda (F)
31	1	E. T. Byram
34	1	C. S. Bowyer
35	1	F. D. Seward
36	1	T. A. Chubb
42	1	H. D. Tananbaum
43	1	E. M. Kellogg
54	1	H. Friedman
74	1	R. Giacconi
81	1	H. Gursky

Note: All of the above scientists are experimentalists except those marked 'T' (for theoretician) and 'O' (for optical, radio or infrared astronomer). All are from the United States except for those marked 'F' (for foreigner).

Data compiled from *Astronomischer Jahresbericht* (Berlin: Verlag W. de Gruyter, published annually between 1962 and 1968), and *Astronomy and Astrophysics Abstracts* (Berlin: Springer-Verlag, published biennially between 1969 and 1972).

Summaries of Appendix 1

	Number of scientists	Percentage of total
Type of scientists		
X-ray experimentalists	98	77.8
Ground-based astronomers	14	11.1
Theoreticians	14	11.1
Total	126	100.0
National origin of scientists		
United States	91	72.2
India	9	7.1
Japan	9	7.1
Australia	8	6.3
Great Britain	3	2.4
Italy	3	2.4
Netherlands	1	0.8
Canada	1	0.8
Soviet Union	1	0.8
Total	126	100.0

Data compiled from *Astronomischer Jahresbericht* (Berlin: Verlag W. de Gruyter, published annually between 1962 and 1968), and *Astronomy and Astrophysics Abstracts* (Berlin: Springer-Verlag, published biennially between 1969 and 1972).

Experimental groups in nonsolar X-ray astronomy 1960–72

Group	Vehicles used	Source of funding	NASA funding 1960–72 ($000)
Adelaide and Tasmania, Univs. (Australia)	balloons, rockets	NASA UKSRC AFOSR, ARGC	460
Am. Sci. & Eng.	rockets, satellite	AFOSR, NASA	16246
Bologna, Univ. (Italy)	balloons rockets	INRC	
Calgary, Univ. (Canada)	rockets	NRCC	
Ca., Berkeley, Univ.	rockets	NASA	978
Ca., San Diego, Univ.	balloons, satellite	NASA	3723 (a)
Columbia Univ.	rockets	NASA, AFOSR	3212 (a)
Goddard Space Flight Center	balloons, rockets	NASA	(b)
Lawrence Rad. Lab.	rockets	AEC	
Leicester Univ. (UK)	rockets	UKSRC, NASA	
Lockheed Miss. & Space Co.	rockets satellite	NASA	2180
Los Alamos Sci. Lab.	satellites	AEC	
Mass. Inst. of Tech.	balloons, rockets	NASA	12099 (a)
Nagoya, Univ. (Japan)	rockets	ISAS	
Nav. Res. Lab.	rockets	ONR, NASA	21763 (a, c)
Phys. Res. Lab. Ahmedabad (India)	rockets	NASA	
Rice Univ.	balloons	AFOSR	

Appendix (*cont.*)

Group	Vehicles used	Source of funding	NASA funding 1960–72 ($000)
Royal Dutch Acad., Delft (Netherlands)	balloons	Dutch Meteor. Inst.	
Tata Inst. of Fund. Res. (India)	balloons	Balloon Flts. Grp. of Tata Inst.	
Tokyo, Univ. (Japan)	balloons	ISAS	
Wisconsin, Univ.	rockets	NASA	3265 (*a*)

Abbreviations:

NASA = National Aeronautics and Space Administration
UKSRC = United Kingdom Science Research Council
AFOSR = Air Force Office of Scientific Research
ARGC = Australian Research Grants Committee
INRC = Italian National Research Council
NRCC = National Research Council of Canada
AEC = Atomic Energy Commission
ISAS = Institute of Space and Aeronautical Science, Tokyo
ONR = Office of Naval Research

Notes:
(*a*) This sum includes grant money for research in broad fields of space astronomy and not just nonsolar X-ray astronomy.
(*b*) Though affiliated with NASA, the Goddard group competed for grants like outside teams. Nevertheless, funding information was difficult to obtain.
(*c*) This sum also includes grant money for research in solar X-ray astronomy.

APPENDIX 3

Technical discussion

Units

X-rays form part of the electromagnetic spectrum, which ranges from the low-energy radio region to the high-energy gamma ray region. Within the X-ray region itself, investigators often distinguish between less energetic 'soft' X-rays (about 0.1 keV) and more powerful 'hard' X-rays (up to 1000 keV). Like all electromagnetic radiation, X-rays can also be described by their wavelength (λ, in angstroms): soft X-rays have long wavelengths (up to 100 Å) while hard X-rays have short wavelengths (less than 1 Å). To determine an X-ray photon's wavelength from its energy, one uses the formula:

$$\lambda \text{ (in angströms)} = 12.4/\text{energy (in keV)}.$$

Although kilo-electron volts and angströms are the two most commonly used units for describing X-rays, one can also describe the frequency of the radiation (ν, in hertz or cycles s^{-1}), as one does in radio astronomy. The conversion relationship is:

$$\nu = 2.418 \times 10^{17} \times \text{energy (in keV)}.$$

To summarize with an example, a 10 keV X-ray photon has a corresponding wavelength of 1.24 Å and a frequency of 2.418×10^{18} hertz.

Detectors above the Earth's atmosphere 'see' X-rays through their metal or plastic windows. The intensity or flux observed is described in terms of the number of photons passing through a unit area of detector window in a unit time. Normally, one writes this as photons cm^{-2} s^{-1}. Because X-ray photons have energy associated with them, flux can also be expressed in erg cm^{-2} s^{-1}. From rockets, instruments observe the Sun's X-ray flux to be about 10^6 photons cm^{-2} s^{-1} or 1 erg cm^{-2} s^{-1} in wavelengths shorter than about 105 Å. In contrast to this huge intensity, the flux from Sco X-1 is only 28 photons cm^{-2} s^{-1} (a revised figure since the first detection of about one-sixth this value) or 4×10^{-7} erg cm^{-2} s^{-1} near the peak wavelength of 3 Å. Although it appears much dimmer than the Sun, Sco X-1 has a large intrinsic luminosity, or energy output, as can be determined by measuring its flux and using the relationship:

$$\text{luminosity} = \text{flux} \times 4\,\pi \times (\text{distance})^2.$$

Luminosity is usually expressed in units of erg s^{-1}. At 10^{36} erg s^{-1}, the X-ray luminosity of Sco X-1 is about 10^6 times greater than the Sun's. The X-ray source only looks like a weak emitter because of its great distance from the Earth.

Detectors

Proportional counters, Geiger counters, and scintillation counters constitute the most common types of detectors used in X-ray astronomy. All work on the same principle: when an X-ray photon interacts with matter in the instrument, it creates a cascade of electrons which is amplified electronically and recorded. Proportional and Geiger counters are basically the same instruments: shallow boxes made of electrically conducting materials and covered by a thin sheet of metal or plastic that permit the passage of X-rays (Figure 27). Inside are gases, mostly inert elements such as argon or xenon, and an anode wire, which is stretched across the vessel and connected to an electronic amplifier. If the anode is maintained at a potential of about 2000 volts, the unit becomes a proportional counter. When an X-ray photon enters the vessel, a gas atom absorbs it, releasing an electron whose energy is proportional to the incident radiation. The electron ionizes other atoms and liberates still more that are drawn to the positively charged anode to produce an electrical pulse. Each pulse lasts about 10^{-6} seconds and corresponds to one X-ray photon. Because all the electrons have energies proportional to the X-ray photon, the output pulse magnitude is also proportional to the original photon's. Using appropriate electronics, the energy of the X-ray can be determined and the radiation's spectrum can be recorded.

When scientists increase the anode's potential above 2000 volts, the device measures only the number of photons striking it. Known as a Geiger counter, the instrument, whose use dates back to the 1930s, cannot yield information on the energies of X-rays. However, it produces larger and more easily recognized signals from incident radiation while using simpler electronics. To obtain a crude estimate of an X-ray spectrum, scientists can simultaneously use several Geiger counters having various window materials. Each film covering will absorb radiation of different wavelengths, thus permitting through only photons of known energies. In X-ray astronomy's history, the easily made Geiger counters preceeded proportional counters. Both counters' sensitivity depends on many factors, such as the gases used in the vessel. But perhaps most important is the size of the window. A large aperture 'captures' more X-ray photons than a small one.

Because the gases in proportional and Geiger counters do not respond well to high-energy X-rays (greater than 20 keV), experimentalists use semiconductor crystals in a scintillation counter. As X-ray photons strike the crystal, they release electrons and produce small flashes of light. Amplified by a photomultiplier, the light pulses are proportional to the photon's energy (Figure 28). Just like proportional counters, scintillation counters provide energy spectra.

Cosmic ray particles complicate the use of all these instruments by producing electron cascades when they interact with gases or crystals. To help eliminate the problem, investigators place magnets above the windows to sweep away charged cosmic ray particles before entering. Better still, they use anticoincident counters, which consist of X-ray and cosmic ray detectors joined together. Electronic logic systems distinguishes between X-ray photons, which cannot penetrate the metal walls of the cosmic ray detector, and material cosmic ray particles. In the late 1960s, pulse shape discrimination became a useful way to distinguish electronically between short-lived X-ray pulses and longer cosmic ray pulses in detectors (Figure 20). Commonly, experimentalists use a combination of these techniques to avoid cosmic ray contamination during X-ray observations.

Collimators

To determine the angular size and location of X-ray sources, scientists employ collimators. Usually made of parallel metal plates arranged in rectangles (Figure 14), collimators are placed in front of

Figure 27. A schematic diagram of the type of proportional counter used in X-ray astronomy. From Longair, *High Energy Astrophysics*, p. 73.

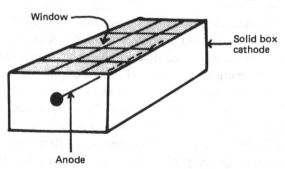

Window

Solid box
cathode

Anode

detector windows to restrict the field of view, just as a long cylinder (with or without lenses in it) restricts a person's view when held to the eye. Passing across an X-ray source, the detector notes a radiation peak. Because the collimator defines a finite solid angle viewed at each moment, a scan of a source identifies a narrow path in which the object lies. If the detector passes the same source a second time, the intersecting paths define a small region in which the body lies. Narrow-angled collimators provide the finest position determinations, but they do not permit wide-angle searches for new sources. Using an uncollimated detector for such a survey, the AS&E scientists discovered the first X-ray source in 1962, but they could not determine its location precisely. Looking specifically for the bright object in 1963, the NRL team used collimated detectors that restricted the view to a few degrees, thus making possible more accurate measurements of the source's angular size and location.

Modulation collimators provide better determinations of an object's angular size than do simple baffle systems (Figures 15 and 16). Consisting

Figure 28. A scintillation counter showing the large scintillating crystal and the photomultiplier tube (RCA 7046). From Longair, *High Energy Astrophysics*, p. 78.

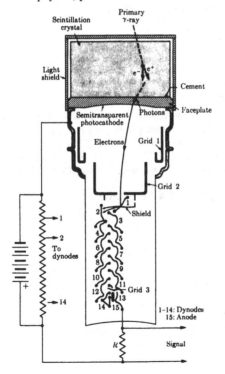

of two or more parallel wire grids, the collimator is placed in front of the detector window, but it determines the angular size of sources from the shadows cast on the counter. Assuming a simple collimator having two grids, separated by a distance D, one can imagine what happens when the system observes small and extended sources. If detecting radiation from a point source with angular diameter γ, (γ being much less than d/D, where d is the diameter of the wires), then the first grid produces parallel shadows that would fall on the second grid and detector. Rotating past a source, the collimator alternately transmits and obstructs (i.e., modulates) the incident radiation. However, if the source has a large angular diameter, with γ greater than d/D radians, the projected rays do not arrive parallel to each other. The shadow cast by the first grid is not clear and the flux is not modulated. Hence, the degree of modulation indicates the source's angular size. Because diffraction effects are negligible (since X-ray wavelengths are much shorter than the diameter of the wires), scientists can make high-resolution determinations of angular size. Using a four-grid modulation collimator in 1966, the AS&E-MIT groups determined Sco X-1's angular size to be smaller than 20 arc seconds.

Table 5. *Assumed radiation mechanisms and spectra of X-ray sources*

Radiation mechanism	Spectral proportionalities
Synchrotron, Inverse Compton ('Power Law')	$I(v) \propto v^{-(a)}$
Bremsstrahlung	$I(v) \propto \exp(-hv/kT)$
Blackbody	$I(v) \propto v^3/[\exp(hv/kT)-1]$

I = intensity v = frequency a = dimensionless number; h = Planck's constant; k = Boltzmann's constant; T = temperature (in degrees Kelvin).
For more details, see Wallace H. Tucker, *Radiation Processes in Astrophysics* (Cambridge, Massachusetts: MIT Press, 1975), and M. S. Longair, *High Energy Astrophysics* (Cambridge, England: Cambridge University Press, 1981).

Figure 29. Assumed spectra of X-ray sources.

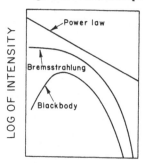

LOG OF PHOTON ENERGY

X-ray absorption

The familiar diagnostic use of X-rays for observing bones through flesh implies to most people that the radiation is very penetrating. Still, it only takes a few centimeters of matter to absorb X-rays and create ionized atoms. The expression for their absorption is:

$$I = I_0 \exp(-\mu x)$$

where I_0 = incident flux; I = flux remaining after passing through material; x = distance traversed by radiation; and μ = mass absorption coefficient. This coefficient can be further analyzed as:

$$\mu = \sigma N_0 / A \ cm^2 \ g^{-1}$$

where σ is the photoelectric cross section of the absorbing material (in cm^2); N_0 is Avogadro's number; and A is the atomic weight of the material. The high absorption coefficient of air in the atmosphere's E-region (at a height of about 100 km) explains why X-ray astronomy observations must be performed in the near vacuum above it. Furthermore, the cross section in the energy range of 1–10 keV varies approximately as the fourth power of atomic number, Z. This fact explains why experimentalists used low-Z beryllium as a window material (for minimal absorption) and high-Z gases to absorb photons in counters.

X-Ray Spectra

The energy spectra of X-ray sources can tell much about the way the radiation is produced. For example, if a bremsstrahlung spectrum is observed from a source, one knows that thin plasmas at high temperatures exist. Table 5 lists the basic relationships between the intensity and frequency of radiations that characterize the various spectra. Figure 29 provides a graphical representation of the spectra.

Bibliographic note on unpublished sources

A large amount of information used in this book came from unpublished documents. Listed below are the most important sources of these materials and brief descriptions of them.

    I. National Aeronautics and Space Administration Headquarters, Washington, D.C.

        1. Office of Space Science. Files of research proposals; reviews of proposals; and final reports of contracts. (Several final reports are abstracted in *Scientific and Technical Aerospace Reports*, published by NASA.)

        2. Office of Procurement. Information on contracts and grants.

        3. Office of University Affairs. Detailed information on funding for university groups.

        4. History Office. General archives; biographical files; project files; papers of Homer E. Newell, Jr; transcribed interviews of General Andrew J. Goodpaster, Jr, James R. Killian, Jr, and Edward Purcell.

    II. Smithsonian Institution, Washington, D.C.

        1. National Air and Space Museum Library. Files on specific research vehicles; excellent file on the response to the Sputnik crisis of 1957, including United Press International wire service reports.

        2. Smithsonian Archives. Papers of Riccardo Giacconi (from 1959 to about 1975).

    III. National Academy of Sciences, Washington, D.C.
Space Science Board. Minutes of the Board's committee meetings from 1958 to the present.

    IV. Naval Research Laboratory, Washington, D.C.
Historian's Office. Organizational charts; manuscript histories of laboratory research; reports of rocket experiments; and transcribed interview of Edward O. Hulburt.

    V. Institute for Scientific Information, Philadelphia, Pennsylvania.
Citation data derived from the magnetic tape version of the *Science Citation Index*, 1970–7.

VI. American Institute of Physics, New York City, New York.
Transcribed interviews with astrophysicists, including those of Subrahmanyan Chandrasekhar, William A. Fowler, and Thomas Gold.

VII. Interviews with Scientists, Engineers, and Administrators performed by the author.
(Several interviewees gave the author unpublished materials from their files to support statements made during our discussions. The affiliations listed were those when interviewed. Interviews marked with an asterisk are transcribed and available through the American Institute of Physics, Center for History of Physics, New York City.)
P. Frank Winkler, Middlebury College, Middlebury, Vermont and MIT, Cambridge, Massachusetts, 6 June 1976 and 5 July 1976.
Philip C. Fisher, Lockheed Missiles and Space Company, Palo Alto, California, 6 June 1976.
Robert Bless, University of Wisconsin, Madison, Wisconsin, 11 June 1976.
Homer E. Newell, Jr, NASA Headquarters, Washington, D.C., 17 June 1976, 28 June 1976, 12 September 1977, 31 January 1979, 17 July 1980,* and 20 October 1980.*
James E. Kupperian, Jr, NASA Goddard Space Flight Center, Greenbelt, Maryland, 25 June 1976 and 12 September 1977.
Talbot A. Chubb, NRL, Washington, D.C., 28 June 1976, 22 July 1976, 2 May 1978.
Malcolm P. Savedoff, University of Rochester, Rochester, New York, 29 June 1976.
Nancy G. Roman, NASA Headquarters, 30 June 1976.
Giovanni G. Fazio, Smithsonian Astrophysical Observatory, Cambridge, Massachusetts, 8 July 1976.
Arthur D. Code, University of Wisconsin, 9 July 1976.
Riccardo Giacconi, Smithsonian Astrophysical Observatory, 12 July 1976, 13 February 1979, and 15 July 1980.
George W. Clark, MIT, 15 July 1976, 20 October 1978, and 14 February 1979.
Bruno B. Rossi, MIT, 21 July 1976 and 14 February 1979.
Martin Annis, AS&E, Cambridge, Massachusetts, 22 July 1976 and 12 February 1979.
Edward T. Byram, NRL, Washington, D.C., 22 July 1976.
John W. Salisbury, Energy Research and Development Administration, Washington, D.C., 27 July 1976.
C. Stuart Bowyer, University of California, Berkeley, California, 28 July 1976 and 10 March 1978.
Alan Bunner, University of Wisconsin, 24 November 1976.
Jack A. Matice, Aeroject Liquid Rocket Company, Sacramento, California, 20 July 1977.
Al M. Olson, Aerojet Liquid Rocket Company, 20 July 1977.
S. Fred Singer, University of Virginia, Charlottesville, Virginia, 25 July 1977.
William A. Russell, Jr, NASA Goddard Space Flight Center, 1 August 1977.
George Kraft, NASA Goddard Space Flight Center, 1 August 1977.
Jon Busse, NASA Goddard Space Flight Center, 1 August 1977.

David Shrewsberry, NASA Goddard Space Flight Center, 9 August 1977.

Albert Boggess, III, NASA Goddard Space Flight Center, 28 February 1978.

Charles D. Swift, Lawrence Livermore Laboratory, Lawrence, California, 9 March 1978 and 1 June 1982.

Roderick J. Grader, Lawrence Livermore Laboratory, 9 March 1978.

Hugh M. Johnson, Lockheed Missiles and Space Company, 10 March 1978.

Loren W. Acton, Lockheed Missiles and Space Company, 10 March 1978 and 4 June 1982.

Richard C. Catura, Lockheed Missiles and Space Company, 10 March 1978.

Frederick D. Seward, Smithsonian Astrophysical Observatory, 27 March 1978.

William Grasberger, Lawrence Livermore Laboratory, 28 March 1978.

Jerry P. Conner, Los Alamos Scientific Laboratory, Los Alamos, New Mexico, 13 April 1978.

Hans Mark, Department of Defense, Washington, D.C., 14 April 1978.

John E. Evans, Lockheed Missiles and Space Company, 1 August 1978.

Martin Walt, Lockheed Missiles and Space Company, 1 August 1978.

Roland E. Meyerott, Lockheed Missiles and Space Company, 2 August 1978.

William L. Kraushaar, University of Wisconsin, 24 October 1978.

James L. Matteson, University of California, San Diego, California, 22 November 1978.

Dan McCammon, University of Wisconsin, 4 December 1978.

Richard C. Henry, Johns Hopkins University, Baltimore, Maryland, 5 December 1978.

Jacob I. Trombka, NASA Goddard Space Flight Center, 7 December 1978.

Herbert Gursky, Smithsonian Astrophysical Observatory, 12 February 1979.

Paul Gorenstein, Smithsonian Astrophysical Observatory, 13 February 1979.

Robert W. Lehnert, LND Inc., Oceanside, New York, 27 March 1979.

Franklin G. Tate, NASA Headquarters, 3 May 1979.

George B. Field, Smithsonian Astrophysical Observatory, 14 and 15 July 1981.*

Herbert Friedman, NRL, 21 August 1981.*

J. Peter Stoering, Lawrence Livermore Laboratory, 1 June 1982.

Richard Rodrigues, Lawrence Livermore Laboratory, 1 June 1982.

Frank R. Harnden, Jr, Smithsonian Astrophysical Observatory, 7 June 1982.

Gerry Chodil, Ball Aerospace Systems Division, Boulder, Colorado, 7 June 1982.

VIII.   Letters from X-ray astronomy investigators to author.

Riccardo Giacconi, Smithsonian Astrophysical Observatory, 14 July 1976, 30 December 1976, and 19 October 1979.

Nancy G. Roman, NASA Headquarters, 27 August 1976.

Herbert Friedman, NRL, 16 September 1976 and 11 September 1980.

Homer E. Newell, Jr, NASA Headquarters, 20 December 1976.

George W. Clark, MIT, 21 September 1977.

Hugh M. Johnson, Lockheed Missiles and Space Company, 14 February 1978.

Roger J. Francey, University of Tasmania, Australia, 10 July 1978.

P.J.N. Davison, University of Adelaide, Australia, 26 July 1978.

Richard M. Thomas, University of Tasmania, 8 August 1978.
Richard C. Henry, Johns Hopkins University, 12 Decmeber 1978.
Ken G. McCracken, University of Adelaide, 4 January 1979.
Satio Hayakawa, Nagoya University, Japan, 6 April 1979.
Ken A. Pounds, University of Leicester, England, 30 April 1979.

NOTES

Introduction

1. For a general introduction on viewing the universe through different windows, see Harry Augensen and Jonathan Woodbury, 'The Electromagnetic Spectrum,' *Astronomy* 10 (No. 6, 1982), pp. 6 – 22.
2. National Science Board, *Science Indicators* 1978, National Science Foundation (Washington, D.C.: Government Printing Office, 1979), pp. 196 and 212. (This volume contains data for years prior to 1978.)
3. Good accounts of the military sponsorship of astronomical research include Sir Bernard Lovell, 'The Effects of Defense Science on the Advance of Astronomy,' *Journal for the History of Astronomy* 8 (1977), pp. 151 – 73; and Martin Harwit, *Cosmic Discovery: The Search, Scope, and Heritage of Astronomy* (New York: Basic Books, 1981), pp. 245 – 6.

Chapter 1: The heritage of X-ray astronomy

1. Clive Hart, *Kites: An Historical Survey* (New York: Frederick A. Praeger, 1967), pp. 81 – 98.
2. Lennart Ege, *Balloons and Airships* (New York: Macmillan, 1974), p. 101; and G. Pfotzer, 'History of the Use of Balloons in Scientific Experiments,' *Space Science Reviews* 13 (1972), p. 203.
3. Antoine-Laurent Lavoisier, *Oeuvres de Lavoisier*, vol. 3, p. 733, cited in Henry Guerlac, *Antoine-Laurent Lavoisier: Chemist and Revolutionary* (New York: Charles Scribner's Sons, 1975), p. 100.
4. Pfotzer, 'History of the Use of Balloons,' pp. 207 – 8.
5. Sir Napier Shaw, *Manual of Meterology*, vol. 1 (Cambridge: Cambridge University Press, 1926), p. 225.
6. Victor F. Hess, 'Ueber Beobachtungen der durchdringenden Strahlung,' *Physikalische Zeitschrift* 14 (1913), pp. 610 – 7.
7. Bruno B. Rossi, *Cosmic Rays* (New York: McGraw-Hill, 1964), p. 59.
8. *Ibid.*, p. 70.
9. I.S. Bowen, R.A. Millikan, and V. Neher, 'The Influence of the Earth's Magnetic Field on Cosmic Ray Intensities Up to the Top of the Atmosphere,' *Physical Review* 52 (1937), pp. 80 – 8; and Arthur H. Compton, 'A Geographic Study of Cosmic Rays,' *Physical Review* 43 (1933), pp. 387 – 403.
10. Pfotzer, 'History of the Use of Balloons,' p. 232; P. Freier, E. J. Lofgreen, E. P. Ney, F. Oppenheimer, H. L. Bradt, and B. Peters, 'Evidence for Heavy Nuclei in the Primary Cosmic Radiation,' *Physical Review* 74 (1948),

pp. 213–7; and P. Freier, E. J. Lofgreen, E. P. Ney, and F. Oppenheimer, 'The Heavy Component of Primary Cosmic Rays,' *Physical Review* 74 (1948), pp. 1818–27.

11. A. E. Kennelly, 'On the Elevation of the Electrically Conducting Strata of the Earth's Atmosphere,' *Electrical World and Engineer* 39 (15 March 1902), p. 473.

12. C. S. Gillmor, 'The History of the Term "Ionosphere".' *Nature* 262 (1976), p. 347.

13. E. V. Appleton and M. A. F. Barnett, 'Local Reflection of Wireless Waves from the Upper Atmosphere,' *Nature* 115 (1925), pp. 333–4; G. Breit and M. A. Tuve, 'A Radio Method of Estimating the Height of the Conducting Layer,' *Nature* 116 (1925), p. 357; and G. Breit and M. A. Tuve, 'A Test of the Existence of the Conducting Layer,' *Physical Review* 28 (1926), pp. 554–75.

14. L. A. Gebhard, *Evolution of Naval Radio Electronics and Contributions of the Naval Research Laboratory* (Washington, D.C.: Department of the Navy, 1976), p. 27.

15. Edward A. Marshall, 'Edison's Plan for Preparedness,' *New York Times Magazine* 64 (30 May 1915), p. 7.

16. David K. Allison, 'The Origin of the Naval Research Laboratory,' *US Naval Institute Proceedings* 105 (July 1979), p. 69.

17. Gebhard, *Evolution of Naval Radio Electronics*, p. 37; and J. A. Sanderson, 'Optics at the Naval Research Laboratory,' *Applied Optics* 6 (1967), p. 2029.

18. For example, the Bureau of Engineering originally supported the research performed in the Radio and Sound Divisions. Allison, 'The Origin of the Naval Research Laboratory,' p. 69.

19. Gebhard, *Evolution of Naval Radio Electronics*, p. 37.

20. E. O. Hulburt and A. H. Taylor, 'The Propagation of Radio Waves over the Earth,' *Physical Review* 27 (1926), pp. 189–215.

21. J. A. Ratcliff and K. Weekes, 'The Ionosphere,' in J. A. Ratcliff, ed., *Physics of the Upper Atmosphere* (New York: Academic Press, 1960), p. 379.

22. S. Chapman, 'The Absorption and Dissociative or Ionizing Effect of Monochromatic Radiation on an Atmosphere on a Rotating Earth,' *Proceedings of the Physical Society* 43 (1931), pp. 26–45, and 483–501; and S. Chapman, 'Some Phenomena of the Upper Atmosphere,' *Proceedings of the Royal Society* 132A (1931), pp. 354–6.

23. See, for example, E. O. Hulburt, 'Ionization in the Upper Atmosphere of the Earth,' *Physical Review* 31 (1928), pp. 1018–37; E. O. Hulburt, 'Tables of the Ionization in the Upper Atmosphere,' *Physical Review* 39 (1932), pp. 977–92; and D. R. Bates and H. S. W. Massey, 'The Basic Reactions in the Upper Atmosphere, I,' *Proceedings of the Royal Society* 187A (1946), pp. 261–96.

24. E. O. Hulburt, 'Photoelectric Ionization in the Ionosphere,' *Physical Review* 53 (1938), p. 350.

25. L. Vegard, 'Vorgange und Zustand in der Nordlichtregion,' *Geofysiske Publikasjoner* 12 (No. 5, 1938), p. 18; and Hulburt, 'Photoelectric Ionization,' p. 350. Hulburt was less emphatic about the suggestion than Vegard, maintaining also the possibility of neutral corpuscular radiation as the source of E-region ionization.

26. W. Grotrian, 'Ueber die Intensitaetsverteilung des kontinuierlichen Spektrums der inneren Korona,' *Zeitschrift fur Astrophysik* 3 (1932), pp. 219–22.

27. B. Lyot, 'Quelques Observations de la Couronne Solaire et des Protuberances en 1935,' *L'Astronomie* 51 (1937), p. 215.

28. Bengt Edlen, 'Die Deutung der Emissionslinien im Spektrum der Sonnenkorona,' *Zeitschrift fur Astrophysik* 22 (1942), pp. 30–64; Bengt Edlen, 'The Identification of the Coronal Lines,' *Monthly Notices of the Royal*

Astronomical Society 105 (1945), pp. 323–33. For a review of Edlen's work, see P. Swings, 'The Line Spectrum of the Solar Corona,' *Publications of the Astronomical Society of the Pacific* 57 (1945), pp. 117–37.

29. Frank H. Winter, 'Birth of the VfR: The Start of Modern Astronautics,' *Spaceflight* 19 (1977), p. 243.

30. Wernher von Braun and Frederick I. Ordway, III, *History of Rocketry and Space Travel*, 3rd edn. (New York: Thomas Y. Crowell, 1975), p. 108.

31. Walter R. Dornberger, 'The German V-2,' in Eugene M. Emme, ed., *The History of Rocket Technology* (Detroit: Wayne State University Press, 1964), p. 29.

32. William R. Corliss, *NASA Sounding Rockets, 1958–68: A Historical Summary* (Washington, D.C.: NASA, 1971), p. 12.

33. L. D. White, 'Project Hermes V-2 Missile Program, 'General Electric Company report R52A0510, September 1952, p. 3.

34. E. Durand, 'Rocket Sonde Research at the Naval Research Laboratory,' in Gerard P. Kuiper, ed., *The Atmospheres of the Earth and Planets* (Chicago: University of Chicago Press, 1949), p. 135.

35. Milton W. Rosen, *The Viking Rocket Story* (New York: Harper and Brothers, 1955), p. 23.

36. Daniel J. Kevles, 'Scientists, the Military, and the Control of Postwar Defense Research: The Case of the Research Board for National Security, 1944–46,' *Technology and Culture* 16 (1975), p. 20; The Bird Dogs, 'The Evolution of the Office of Naval Research,' *Physics Today* 14 (No. 8, 1961), p. 30; and Daniel S. Greenberg, *The Politics of Pure Science*, revised edn. (New York: New American Library, 1971), p. 133; John E. Pfeifer, 'The Office of Naval Research,' *Scientific American* 180 (No. 2, 1949), pp. 11 and 14; and Adam Yarmolinsky, 'Military Sponsorship of Science and Research,' in *The Military Establishment: Its Impact on American Society* (New York: Harper and Row, 1971), p. 291.

37. Interview with Talbot A. Chubb, 2 May 1978.

38. Herbert Friedman, 'Rocket Astronomy,' *Annals of the New York Academy of Sciences* 198 (1972), p. 267.

39. W. A. Baum, F. S. Johnson, J. J. Oberly, C. C. Rockwood, C. V. Strain, and R. Tousey, 'Solar Ultraviolet Spectrum to 88 Kilometers,' *Physical Review* 70 (1946), p. 781; and E. Durand, J. J. Oberly, and R. Tousey, 'Solar Absorption Lines Between 2950 and 2200 Angströms,' *Physical Review* 71 (1947), p. 827. Although reaching an altitude of 88 km, the rocket returned useful spectra only up to a height of 55 km.

40. Leo Goldberg, 'Astronomy from Artificial Satellites,' *Annual Report of the Board of Regents of the Smithsonian Institution for the Year* 1959 (1960), pp. 288–9.

41. T. R. Burnight, 'Soft X-Radiation in the Upper Atmosphere,' *Physical Review* 76 (1949), p. 165; and T. Robert Burnight, 'Ultraviolet Radiation and X-Rays of Solar Origin,' in C. S. White and O. O. Benson, Jr, *Physics and Medicine in the Upper Atmosphere* (Albuquerque: University of New Mexico Press, 1952), p. 231.

42. Theodore Lyman, 'The Transparency of the Air Between 1100 and 1300 A,' *Physical Review* 48 (1935), pp. 149–51.

43. J. D. Purcell, R. Tousey, and K. Watanabe, 'Observations at High Altitudes of Extreme Ultraviolet and X-Rays from the Sun,' *Physical Review* 76 (1949), pp. 165–6.

44. R. Tousey, K. Watanabe, and J. D. Purcell, 'Measurements of Solar Extreme Ultraviolet and X-Rays from Rockets by Means of a $CaSO_4$:Mn Phosphor,' *Physical Review* 83 (1951), p. 795.

45. These experiments appeared to have suffered from interpretative problems as well. We now know that the beryllium filters used by Burnight and Tousey may have been too thick to permit X-rays to penetrate to the plates. In Burnight's experiments, the varying heat and pressure during the rocket flight may have caused the observed exposure. Interview with Talbot A. Chubb, 2 May 1978.

46. Interview with Herbert Friedman, 21 August 1980.

47. Results of Friedman's dissertation work were published in W. W. Beeman and H. Friedman, 'The X-Ray K Absorption Edges of the Elements Fe (26) to Ge (32),' *Physical Review* 56 (1939), pp. 392–405; J. A. Bearden and H. Friedman, 'The X-Ray K beta 2, 5-Emission Lines and K-Absorption Limits of Cu – Zn Alloys,' *Physical Review* 58 (1940), pp. 387 – 95; and H. Friedman and W. W. Beeman, 'Copper and Nickel X-Ray K beta 2 and K beta 5-Emission Lines and K-Absorption Limits in Cu – Ni Alloys,' *Physical Review* 58 (1940), pp. 400 – 6.

48. H. Friedman, 'Geiger Counter Spectrometer for Industrial Research,' *Electronics* 18 (1945), p. 132.

49. 'Herbert Friedman,' in Shirley Thomas, *Men of Space*, vol. 7 (Philadelphia: Chilton Books, 1965), pp. 57 – 8.

50. These counters are described in C. O. Muelhause and H. Friedman, 'Measurement of Betatron Radiation with G – M Counters,' *Physical Review* 69 (1946), pp. 691 – 2; C. O. Muelhause and H. Friedman, 'Geiger – Muller Counter Technique for High Counting Rates,' *Physical Review* 69 (1946), p. 46; C. O. Muelhause and H. Friedman, 'Applications of Electronic Methods of the Measurement of X-ray and Gamma-Ray Intensities,' *Industrial Radiography* 6 (1947), pp. 9 – 20; H. Friedman and L. S. Birks, 'A Geiger Counter Spectrometer for X-Ray Fluorescences Analysis,' *Review of Scientific Instruments* 19 (1948), pp. 303 – 6; and H. Friedman, 'Geiger Counter Tubes,' *Proceedings of the Institute of Radio Engineers* 37 (1949), pp. 791 – 808.

51. H. Friedman, S. W. Lichtman, and E. T. Byram, 'Photon Counter Measurements of Solar X-Rays and Extreme Ultraviolet,' *Physical Review* 83 (1951), p. 1025.

52. See, for example, F. Hoyle and D. R. Bates, 'The Production of the E-Layer,' *Terrestrial Magnetism and Atmospheric Electricity* 53 (1948), pp. 51 – 62.

53. Friedman, 'Photon Counter Measurements,' p. 1025.

54. E. T. Byram, T. A. Chubb, and H. Friedman, 'The Contribution of Solar X-Rays to E-Layer Ionization,' *Physical Review* 92 (1953), pp. 1066 – 7. Also see Gerhard Elwert, 'The Effect of Soft X-Rays from the Solar Corona on the formation of the Normal Ionospheric E-layer,' *Journal of Atmospheric and Terrestrial Physics* 4 (1953), pp. 68 – 77.

55. Interview with Talbot A. Chubb, 2 May 1978.

56. Interview with Edward T. Bryam, 2 July 1976.

57. The team's first orbiting experiment was carried on the Solrad I satellite in 1960. See R. W. Kreplin, T. A. Chubb, and H. Friedman, 'X-Ray and Lyman Alpha Emission from the Sun as Measured from the NRL SR-1 Satellite,' *Journal of Geophysical Research* 67 (1962), pp. 2231 – 53.

58. R. L. Blake, T. A. Chubb, H. Friedman, and A. E. Unzicker, 'Interpretation of X-Ray Photograph of the Sun,' *Astrophysical Journal* 137 (1963), pp. 3 – 15.

59. S. Mandelshtam, B. Vasilyev, J. Voron'ko, I. Tindo, and A. Shurygin, 'Measurements of Solar X-Ray Radiation,' *Space Research* 3 (1963), pp. 822 – 35.

60. M. Waldmeir and H. Muller, 'Die Sonnenstrahlung in Gebiet von $\lambda = 10$ cm,'

Zeitschrift fur Astrophysik 27 (1950), pp. 58–72; and G. Elwert, 'The X-Ray Radiation of the Solar Corona and Hot Coronal Condensations,' in W. J. G. Beynon and G. M. Brown, *Solar Eclipses and the Ionosphere* (London: Pergamon Press, 1956), p. 172.

61. Herbert Friedman, 'Solar Radiation,' *Astronautics* 7 (No. 8, 1962), pp. 14 and 20.

62. G. Elwert, 'The Theory of X-Ray Emission of the Sun,' *Journal of Geophysical Research* 66 (1961), p. 394.

63. Gehard Elwert, 'Die weiche Roentgenstrahlung der ungestorten Sonnenkorona' *Zeitschrift fur Naturforschung* 9A (1954), p. 637.

64. For precise determinations of where rockets pointed during the day, scientists developed a number of techniques, none of which were suitable for nighttime work. See T. A. Ecrgstraln, 'Photography from the V-2 at Altitudes Ranging up to 160 Km,' NRL Upper Atmosphere Research report No. 4, NRL report R-3171, NRL, Washington, D.C., October 1947, pp. 119–30; L. W. Fraser and E. H. Siegler, 'High Altitude Research Using the V-2 Rocket, March 1946 to April 1947,' Bumblebee Series Report No. 81, Johns Hopkins University Applied Physics Laboratory, Silver Spring, Maryland, July 1948, pp. 82–3; and D. M. Packer and R. Tousey, 'A Solar Aspect Indicator for a Rocket,' NRL Upper Atmosphere Research report No. 17, NRL report 4024, NRL, Washington, D.C., September 1952.

65. The Aerobee derives its name from its manufacturer, the *Aero*jet Engineering Company, and the Bumble*bee* series of missiles. L. W. Fraser, 'Aerobee High Altitude Sounding Rocket: Design Construction and Use,' Bumblebee Series report No. 95, Johns Hopkins University Applied Physics Laboratory, Silver Spring, Maryland, December 1948, p. 2.

66. J. A. Van Allen, L. W. Fraser, and J. F. R. Floyd, 'The Aerobee Sounding Rocket – a New Vehicle for Research in the Upper Atmosphere,' *Science* 108 (1948), pp. 746–7; and Homer E. Newell, Jr, 'Characteristics of the High Altitude Rocket and a Research Tool,' in White, *Physics and Medicine of the Upper Atmosphere*, p. 411.

67. J. W. Townsend, E. Pressley, R. Slavin and L. Draff, Jr, 'The Aerobee-Hi,' in Homer E. Newell, Jr, *Sounding Rockets*, (New York: McGraw-Hill, 1959), pp. 75–6; J. W. Townsend, Jr, and R. M. Slavin, 'Aerobee-Hi Development Program, *Jet Propulsion* 27 (1957), pp. 264–5; and C. P. Chalfant, E. S. Thomas, B. A. Thornstensen, 'Development of A Four-Fin Aerobee Sounding Vehicle,' report 1640, Aerojet-General Corporation, El Monte, California, June 1960, p. 6.

68. 'Goddard Space Flight Center Sounding Rocket Flight Compendium,' NASA Goddard Space Flight Center, Greenbelt, Maryland, December 1976.

69. Interview with Jack A. Matice and Alan M. Olson, 20 July 1977. Also see prepared statement by Homer E. Newell, Jr in US Congress, House, Committee on Science and Astronautics, 1966 *NASA Authorization Hearings*, 89th Congress, 1st session, part 3, p. 552.

70. Homer E. Newell, Jr, 'The High Altitude Sounding Rocket,' *Jet Propulsion* 27 (1957), p. 262.

71. Interview with Albert Boggess, III, 28 February 1978. (Boggess was a member of the NRL rocket astronomy group in the late 1950s.) More detailed descriptions of the aspect systems are included in James E. Kupperian, Jr and Robert W. Kreplin, 'Optical Aspect System for Rockets,' *Review of Scientific Instruments* 28 (1957), pp. 14–19; and James E. Kupperian, Jr, Albert Boggess, III, and James E. Milligan, 'Observational Astrophysics from Rockets: I. Nebular Photometry at 1300 A,' *Astrophysical Journal* 128 (1958), p. 456.

72. E. T. Byram, T. A. Chubb, H. Friedman, and J. E. Kupperian, 'Rocket

Observation of Extraterrestrial Far-Ultra-Violet Radiation,' *Astronomical Journal* 62 (1957), p. 9; and E. T. Byram, T. A. Chubb, H. Friedman, and J. E. Kupperian, 'Far Ultraviolet Radiation in the Night Sky,' in M. Zelikoff, ed., *The Threshold of Space* (London: Pergamon Press, 1957), p. 209.

73. H. Friedman, 'Space Spectroscopy at the US Naval Research Laboratory – Photoelectric Photometry,' *Journal of Quantitative Spectroscopy and Radiative Transfer* 2 (1962), pp. 551–6.

74. T. A. Chubb and E. T. Byam, 'Rocket Observations of the Far Ultraviolet Sky,' *Space Research* 3 (1963), pp. 1046–60.

75. Kupperian, 'Observational Astrophysics from Rockets,' p. 459.

76. Herbert Friedman, 'Rocket Astronomy – Window into Space,' *Astronautics* 4 (No. 1, 1959), p. 69.

77. E. T. Byram, T. A. Chubb, and H. Friedman, 'On the Absence of Nebular Glow Around Alpha Virginis in the Far Ultraviolet,' *Astrophysical Journal* 139 (1964), pp. 1135–8; and interview with Talbot A. Chubb, 2 May 1978.

Chapter 2: The political environment

1. D. H. Radler, 'The New Red Moon,' *Horizon* 4 (October 1957), p. 1. (*Horizon* is a publication of the Purdue Research Foundation.) For a hint of the Soviet intentions, see John Hillaby, 'Earth Satellite Sought by Soviet; Russian Expert in Denmark Says Success in 2 Years is "Quite Possible",' *New York Times*, 3 August 1955, p. 8.

2. William Leavitt, 'Much News was Bad News...Reactions to Sputnik,' *Air Force Magazine* 40 (December 1957), pp. 35–9.

3. Peter J. Schenk, 'The Sputnik Pearl Harbor,' *Air Force Magazine* 40 (November 1957), pp. 34–7.

4. The Sputnik satellite enabled scientists to carry out geodetic and upper atmospheric density experiments. For a discussion of the International Geophysical Year and its political and diplomatic impact, see US Congress, House, Committee on International Relations, *Science, Technology and American Diplomacy*, vol. 1, Chapter 5, 'The Political Legacy of the International Geophysical Year,' Committee Print (Washington, D.C.: Government Printing Office, 1977), pp. 295–358.

5. United Press International newspaper service report 7, 5 October 1957.

6. '6 out of 7,' *Newsweek*, 28 October 1957, p. 30.

7. United Press International newspaper service report 9, 5 October 1957.

8. 'Around the World in 96 Minutes – What Does it Really Mean?' *US News and World Report*, 18 October 1957, pp. 31–2; and James R. Killian, Jr, *Sputnik, Scientists and Eisenhower* (Cambridge: MIT Press, 1977), p. 3.

9. Statement by Senator Henry Jackson of Washington, in United Press International newspaper service report 91, 5 October 1957.

10. *Kiplinger Washington Letter*, 12 October 1957, p. 2.

11. United Press International newspaper service report 90, 5 October 1957.

12. 'The US Soviet Prestige Race,' *Newsweek*, 2 December 1957, p. 30.

13. 'Missile Miscellany,' *Missiles and Rockets* 2 (November 1957), p. 132.

14. United Press International newspaper service report 36, 5 October 1957.

15. United Press International newspaper service report 85, 5 October 1957.

16. Constance M. Green and Milton Lomask, *Vanguard: A History* (Washington, D.C.: Smithsonian Institution Press, 1971), pp. 189, 197–8.

17. The political debate leading to the creation of NASA is discussed in detail in Enid Curtis Bok Schoettle, 'The Establishment of NASA,' in Sanford A. Lakoff, *Knowledge and Power: Essays on Science and Government* (New York:

Free Press, 1966), pp. 162–270. For an insider's view of the debate, see Homer E. Newell, Jr, *Beyond the Atmosphere: Early Years of Space Science* (Washington, D.C.: Government Printing Office, 1980), pp. 87–99.

18. The story of the US Army's successful launch of the United States' first satellite is contained in John B. Medaris, *Countdown for Decision* (New York: Putnam, 1960).

19. Newell, *Beyond the Atmosphere*, p. 90.

20. 'The US, Ike, and Sputnik,' *Newsweek*, 28 October 1957, p. 31; and *Kiplinger Washington Letter*, 12 October 1957, p. 2.

21. Robert L. Rosholt, *An Administrative History of NASA, 1958–62* (Washington, D.C.: NASA, 1966), pp. 20–1.

22. 'Which Way to Space?' *Washington Post*, 16 June 1958, p. A10.

23. Killian, *Sputnik, Scientists and Eisenhower*, p. 135.

24. Schoettle, 'The Establishment of NASA,' p. 249.

25. National Aeronautics and Space Act of 1958, Public Law 85-568, 85th Congress, HR 12575, 29 July 1958, 72 Stat. 426.

26. Newell, *Beyond the Atmosphere*, p. 98.

27. Rosholt, *An Administrative History of NASA*, p. 47; and Alfred Rosenthal, *Venture into Space: Early Years of the Goddard Space Flight Center* (Washington, D.C.: NASA, 1968), p. 27.

28. Daniel J. Kevles, *The Physicists: The History of a Scientific Community in Modern America* (New York: Alfred A. Knopf, 1978), p. 138.

29. Schoettle, 'The Establishment of NASA,' p. 213.

30. Eli Ginzberg, James W. Kuhn, Jerome Schnee, and Boris Yavitz, 'Transformation of a Science: NASA's Impact on Astronomy,' in *Economic Impact of Large Public Programs: The NASA Experience* (Salt Lake City: Olympus, 1976), p. 87.

31. An exception was Princeton University's Lyman Spitzer, who noted as early as 1940 the advantages of placing optical telescopes in space. *Ibid.*

32. Interview with Arthur D. Code, 9 July 1976. Code himself experienced many of these problems while developing a stellar ultraviolet experiment for the first Orbiting Astronomical Observatory. He began work on the project in 1960, but the launch of the satellite was delayed until 8 April 1966. After being in orbit for two days, the vehicle failed to operate properly, and no data were returned to earth. See George Alexander, 'OAO Fails on Second Day in Orbit,' *Aviation Week and Space Technology* 84 (18 April 1966), p. 31.

33. Interviews with Arthur D. Code, 9 July 1976; and George W. Clark, 15 July 1976. Also see Ginzberg, 'Transformation of a Science,' p. 84.

34. Newell, *Beyond the Atmosphere*, p. 130.

35. Some of the 'pep talks' given at these symposia were published. See Homer E. Newell, Jr, 'Capabilities for Space Research,' *Journal of Geophysical Research* 64 (1959), pp. 1695–712; and Robert Jastrow, 'Symposium on the Exploration of Space, Introductory Remarks,' *Journal of Geophysical Research* 64 (1959), pp. 1647–51.

36. Lloyd V. Berkner and Hugh Odishaw, 'A Note on the Space Science Board,' in Lloyd V. Berkner and Hugh Odishaw, *Science in Space* (New York: McGraw-Hill, 1961), p. 429.

37. National Academy of Sciences, *The National Academy of Sciences: First Hundred Years* (Washington, D.C.: National Academy of Sciences, 1977), p. 555; and minutes of the first meeting of the Space Science Board, Rockefeller Institute, New York City, New York, 27 June 1958.

38. 'Research in Space,' *Science* 130 (1959), p. 201.

39. National Academy of Sciences, *The National Academy of Sciences*, p. 555.

40. 'Interim Report of the Space Science Board Committee on Physics of Fields and Particles in Space,' National Academy of Sciences Space Science Board, Washington, D.C., 24 October 1958, p. 5.

41. Lawrence H. Aller, 'Some Aspects of Ultraviolet Satellite Spectroscopy,' *Publications of the Astronomical Society of the Pacific* 71 (1959), pp. 324–9; Leo Goldberg, 'Summary of Report from the Committee on Optical and Radio Astronomy,' National Academy of Sciences Space Science Board, Washington, D.C., 23–4 October 1959, p. 5; and Leo Goldberg, 'Galactic and Extragalactic Astronomy,' in Berkner, *Science in Space*, pp. 373–6.

42. Such a calculation was performed by Robert Davis in Arthur I. Berman, ed., 'Proceedings of the Conference on X-Ray Astronomy at the Smithsonian Astrophysical Observatory,' Smithsonian Astrophysical Observatory, Cambridge, Massachusetts, 20 May 1960, p. 9.

43. Goldberg, 'Summary of Report,' p. 35. For a retrospective view of the situation by British scientists who later became involved in X-ray astronomy, see J. Leonard Culhane and Peter W. Sanford, *X-Ray Astronomy* (New York: Charles Scribner's Sons, 1981), p. 23.

44. William Grasberger and Louis Henyey, 'Preliminary Report on Interplanetary X-Ray Radiation from Stellar Sources,' Lawrence Radiation Laboratory, Livermore, California, 22 February 1959; and interview with William Grasberger, 24 March 1978.

45. Grasberger, 'Preliminary Report,' pp. II-2 to II-18.

46. T. A. Chubb, H. Friedman, R. W. Kreplin, and J. E. Kupperian, Jr, 'Lyman Alpha and X-Ray Emissions During a Small Solar Flare,' *Journal of Geophysical Research* 62 (1957), pp. 389–98.

47. Friedman recounts these observations in Herbert Friedman, 'Rocket Astronomy,' *Annals of the New York Academy of Sciences* 198 (1972), p. 278.

48. Herbert Friedman, 'NRL Equipment and Plans,' in 'NRL Papers Presented at the Tenth International Astrophysical Symposium at Liege Belguim, 11–15 July 1960,' report 5608, NRL, Washington, D.C., 21 July 1961, p. 5.

49. The upper limit was 10^{-8} erg cm^{-2} s^{-1} Å^{-1}. See Herbert Friedman, 'The Sun's Ionizing Radiation,' in J. A. Ratcliffe, ed., *Physics of the Upper Atmosphere* (New York: Academic Press, 1960), p. 202. Also see Herbert Friedman, 'Information about the Gas Density in Space Derived from Radiation Measurements,' paper presented at AGARD Conference, Paris, France, 26–8 May 1959, AGARDograph 42, pp. 11–12; and J. E. Kupperian, Jr and H. Friedman, 'Gamma Ray Intensities at High Altitudes,' Proceedings of the Fifth CSAGI Assembly, Moscow, USSR, 1958, pp. 1–3.

Chapter 3: First fruit

1. Biographical sketch accompanying AS&E proposal to NASA, 'Measurement of Soft X-Rays from a Rocket Platform Above 150 km,' AS&E document ASE-P-26, AS&E, Inc., Cambridge, Massachusetts, 17 February 1960; and interview with Bruno B. Rossi, 21 July 1976.

2. Minutes of first meeting of the *Ad Hoc* Committee on Space Projects, National Academy of Sciences Space Science Board, Washington, D.C., 30 September 1958.

3. Minutes of the second meeting of the Committee on Space Projects, National Academy of Sciences Space Science Board, Washington, D.C., 11 December 1958.

4. *Ibid.*, attachment 1.

5. Jay Holmes, 'Solar Wind Existence Proven,' *Missiles and Rockets* 8 (29 April 1961), p. 17.

6. Interview with Bruno B. Rossi, 21 July 1976.
7. Interviews with Martin Annis, 22 July 1976 and 13 February 1979; and 'Annual Report, 1967,' AS&E, Inc., Cambridge, Massachusetts, 1967.
8. This philosophy appears to have paid off well. Between 1958 and 1970 the company's sales of goods and scientific services increased 100-fold, from $125000 to $12500000. Among the more successful 'spinoffs' from government-supported research in X-ray astronomy were the low dosage X-ray detectors that are currently used in airport security systems. 'Annual Report, 1970,' AS&E, Inc., Cambridge, Massachusetts, 1970.
9. Letter from Riccardo Giacconi to Joachim Truemper, Max Planck Institute for Extraterrestrial Physics, 3 April 1979.
10. Riccardo Giacconi, George W. Clark, Bruno B. Rossi, 'A Brief Review of Experimental and Theoretical Progress in X-Ray Astronomy,' AS&E document ASE-TN-49, AS&E, Inc., Cambridge, Massachusetts, 15 January 1960.
11. *Ibid.*, pp. 4–6.
12. *Ibid.*, pp. 9–11.
13. Herbert Friedman, 'Rocket Astronomy,' *Scientific American* 200 (No. 6, 1959), p. 59. Also see Herbert Friedman, 'Electromagnetic Radiation from Extra-Terrestrial Sources,' *American Journal of Physics* 28 (1960), p. 626.
14. Riccardo Giacconi, 'Introduction,' in Riccardo Giacconi and Herbert Gursky, eds., *X-Ray Astronomy* (Dordrecht, Holland: D. Reidel, 1974), p. 6.
15. Bruno B. Rossi, 'X-Ray Astronomy,' *Proceedings of the Academy of Arts and Sciences* 106 (No. 4, 1977), p. 39.
16. P. Kirkpatrick and H. H. Pattee, Jr, 'X-Ray Microscopy,' in S. Flugge, ed., *Encyclopedia of Physics*, vol. 30 (Berlin: Springer-Verlag, 1957), p. 305.
17. Hans Wolter, 'Spiegelsystem streifenden Einfalls als abbildende Optiken fur Roentgenstrahlen,' *Annalen der Physik* 10 (1952), pp. 94–114. For a good description of the principles of X-ray reflection techniques, see. James H. Underwood, 'X-Ray Optics,' *American Scientist* 66 (1978), pp. 476–86.
18. Giacconi, 'Introduction,' p. 6.
19. Riccardo Giacconi and Bruno Rossi, 'A "Telescope" for Soft X-Ray Astronomy,' *Journal of Geophysical Research* 65 (1960), pp. 773–5.
20. *Ibid.*; and letter from Riccardo Giacconi to Richard F. Hirsh, 30 December 1976.
21. Riccardo Giacconi and Bruno B. Rossi, 'X-Ray Reflection Collimator Adapted to Focus X-Radiation Directly on a Detector,' Patent 3 143 651 filed 23 February 1961, patented 4 August 1964.
22. Letter from Giacconi to Truemper, 3 April 1979.
23. Baez published reports of what he discussed at the conference in Albert V. Baez, 'A Proposed X-Ray Telescope for the 1–100 Å Range,' *Journal of Geophysical Research* 65 (1960), pp. 3019–20; and Albert V. Baez, 'A Self-Supporting Metal Fresnel Zone-Plate to Focus Extreme Ultraviolet and Soft X-Rays,' *Nature* 186 (1960), p. 958.
24. Conspicuous in their absence at the conference were Herbert Friedman or any member of the NRL group. Their nonattendance did not necessarily indicate a lack of concern in the field. Only six days after the meeting, the scientists performed a rocket experiment at White Sands, New Mexico, and they may have been there a week beforehand. Friedman expressed interest in the conference by permitting portions of his paper concerning a pinhole camera photograph of the Sun to be included along with the transcript of the proceedings. See Arthur I. Berman, ed., 'Proceedings of the Conference on X-Ray Astronomy at the Smithsonian Astrophysical Observatory,'

Smithsonian Astrophysical Observatory, Cambridge, Massachusetts, 20 May 1960.

25. Data provided by NASA Office of Procurement, NASA Headquarters, Washington, D.C.

26. Giacconi, 'Introduction,' p. 7.

27. Giacconi, 'A Brief Review,' US Air Force version, p. 9. Two versions of this AS&E report were printed. The US Air Force version included a discussion concerning lunar X-rays and was sent with proposals to the Cambridge Research Laboratories. The same report, without this discussion, was addressed to NASA with a request for funding of the X-ray telescope.

28. *Ibid.*, p. 10.

29. For example, see Riccardo Giacconi and George Clark, 'Electron – Proton Spectrometer, N-4.' Proposal to the Air Force Cambridge Research Laboratories, AS&E, Inc., Cambridge, Massachusetts, August 1960.

30. Air Force Cambridge Research Laboratories, *Report on Research, July* 1962 *to July* 1963 (Bedford: US Air Force, 1964), p. 6; and Air Force Cambridge Research Laboratories, *Report on Research, July* 1963 *to June* 1965 (Bedford: US Air Force, 1965), p. 7. Also see W. Harry Lambright, *Governing Science and Technology* (New York: Oxford University Press, 1976), pp. 194 – 5.

31. Air Force Cambridge Research Laboratories, *Report on Research, July* 1962 *to July* 1963, p. 5.

32. Interview with John W. Salisbury, 27 July 1976. The US Air Force contract, AF 19 (604)-7214, was referred to in AS&E, 'Measurement of Soft X-Rays from the Moon,' proposal to the Air Force Cambridge Research Laboratories by the AS&E company, AS&E document ASE-83-I, AS&E, Inc., Cambridge, Massachusetts, 25 October 1960, p. 2.

33. 'Rocket-Borne Research,' *History and Progress*, February 1964, p. 5. (*History and Progress* is a publication of the Air Force Cambridge Research Laboratories.)

34. The AS&E project received $210 526 for this research, which began on 11 January 1961 and ended 28 February 1963. The work was performed under US Air Force contract AF 19 (604)-8026. Data provided by AS&E, Inc., Cambridge, Massachusetts.

35. AS&E, 'Measurement of Soft X-Rays from the Moon,' p. 3; interviews with Riccardo Giacconi, 12 July 1976; George W. Clark, 15 July 1976; and Bruno B. Rossi, 21 July 1976.

36. Giacconi, 'Introduction,' p. 8.

37. Air Force Cambridge Research Laboratories, *Report on Research, July* 1962 *to July* 1963, p. 39.

38. Riccardo Giacconi, Herbert Gursky, Frank R. Paolini, and Bruno B. Rossi, 'Evidence for X-Rays from Sources Outside the Solar System,' *Physical Review Letters* 9 (1962), p. 442.

39. *Ibid.*, pp. 439 – 40.

40. *Ibid.*, pp. 441 – 2.

41. Giacconi, 'Introduction,' p. 10.

42. Interview with Herbert Gursky, 12 February 1979.

43. Interview with John D. Salisbury, 27 July 1976.

44. Herbert Gursky, Riccardo Giacconi, Frank R. Paolini, and Bruno B. Rossi, 'Further Evidence for the Existence of Galactic X-Rays,' *Physical Review Letters* 11 (1963), p. 533.

45. *Ibid.*

46. For example, see John Gribbin, *Our Changing Universe: The New Astronomy* (New York: E. P. Dutton, 1976), p. 80.

Chapter 4: Competition and confirmation

1. Interviews with James E. Kupperian, Jr, 22 June 1976 and 25 June 1976.
2. M. P. Savedoff, 'The Crab and Cygnus A as Gamma Ray Sources,' *Nuovo Cimento* 13 (1959), pp. 12–18.
3. Interview with Malcolm P. Savedoff, 29 June 1976; and Malcolm P. Savedoff, 'Far Ultraviolet and Soft X-Ray Astronomy,' proposal for research, in letter to NASA, 3 June 1959.
4. M. P. Savedoff, 'Quarterly Report 3,' 1 July 1961 to 30 June 1962, for NASA contract NSG-32-60, University of Rochester, Rochester, New York, 1962, p. 2.
5. Interview with Giovanni G. Fazio, 8 July 1976.
6. Before 1961, the company was a division of the Lockheed Aircraft Corporation. It then became a subsidiary of the corporation.
7. Interview with Roland E. Meyerott, 2 August 1978. (Meyerott was manager of the Lockheed company's physical sciences laboratory from 1962 to 1966.)
8. Interview with Martin Walt, Lockheed Missiles and Space Company, 1 August 1978. (Walt served as manager of the Lockheed company's physical sciences division since 1966.)
9. Some results of this work appeared in W. L. Imhof, R. V. Smith, and P. C. Fisher, 'Particle Flux Measurements from an Atlas Pod in the Lower Van Allen Belt,' *Space Research* 3 (1963), pp. 438–46; and R. V. Smith, P. C. Fisher, W. L. Imhof, R. D. Moffat, and J. B. Reagan, 'Proton Flux Measurements from Satellites 1961 Sigma-1 and 1961 Alpha-Delta-1 Near the Peak of the Inner Van Allen Belt,' *Space Research* 3 (1963), pp. 463–76.
10. Interview with Philip C. Fisher, 6 June 1976.
11. Philip C. Fisher, 'Proposal for Mapping the Night Sky in the 10 eV to 16 keV Photon Range,' Lockheed document 702 172, Lockheed Missles and Space Company, Palo Alto, California, 12 December 1960, p. 1.
12. Philip C. Fisher and Arthur J. Meyerott, "Stellar X-Ray Emission," *Astrophysical Journal* 139 (1964), p. 124. For more details concerning the detectors, see P. C. Fisher, A. J. Meyerott, H. A. Grench, R. A. Nobles, and J. B. Reagan, 'Soft Particle Detectors,' *IEEE Transactions on Nuclear Science* NS-10 (1963), pp. 211–19.
13. Memorandum from Samuel H. Depew, NASA University Support Programs and Contractor Research Programs, to Paul Gorman, Goddard Space Flight Center Procurement Officer, 18 October 1962; and P. C. Fisher, 'Progress Report for Contract NAS-5-1174,' Lockheed Missiles and Space Company, Palo Alto, California, 10 September 1962, pp. 1–2.
14. In this weapons test on 9 July 1962, a 1.4 megaton nuclear bomb was detonated at an altitude of 400 km. See Wilmot N. Hess, 'The Effects of High Altitude Explosions,' in Donald P. LeGalley and Alan Rosen, eds., *Space Physics* (New York: John Wiley and Sons, 1964), pp. 573–610.
15. C. Stuart Bowyer, 'An Alternate Interpretation of the Paper "Stellar X-Ray Emission," by P. C. Fisher and A. J. Meyerott,' *Astrophysical Journal* 140 (1964), pp. 820–1.
16. Philip C. Fisher and Arthur J. Meyerott, 'Reply to Letter of Stuart Bowyer,' *Astrophysical Journal* 140 (1964), p. 823.
17. Memorandum to Director, NASA Headquarters Contracts Division from Nancy G. Roman, NASA Chief of Astronomy Program, 2 December 1963.
18. Herbert Friedman, 'Rocket Astronomy,' *Scientific American* 200 (No 6, 1959), p. 59; and interview with Herbert Friedman, 21 August 1980.
19. Bruno Rossi, 'X-Ray Astronomy,' *Daedelus* 106 (No. 4, 1977), pp. 40–1.

20. Interview with C. Stuart Bowyer, 28 July 1976.
21. *Ibid.*; and interview with Talbot A. Chubb, 31 May 1978.
22. Charles S. Bowyer, 'Proportional Counter Tube Including a Plurality of Anode–Cathode Units,' Patent 3 383 538, filed 30 December 1965, awarded 14 May 1968.
23. Herbert Friedman, 'X-Ray Astronomy,' *Scientific American* 210 (No. 6, 1964), p. 37.
24. S. Bowyer, E. T. Byram, T. A. Chubb, H. Friedman, 'X-Ray Sources in the Galaxy,' *Nature* 201 (1964), pp. 1307–8.
25. *Ibid.*, p. 1307.
26. *Ibid.*, and S. Bowyer, E. T. Byram, T. A. Chubb, and H. Friedman, 'Galactic X-Ray Astronomy,' *Report of NRL Progress*, February 1964, p. 4.
27. Bowyer, 'X-Ray Sources in the Galaxy,' p. 1308.
28. *Ibid.*
29. 'X-Rays from the Unknown,' *Time*, 1 November 1963, p. 65.
30. The first published report of the NRL observations was printed as an abstract. See S. Bowyer, E. T. Byram, T. A. Chubb, and H. Friedman, 'X-Ray Astronomy,' *Space Research* 4 (1964), p. 966.
31. Walter Sullivan, 'Source of X-Rays Detected in Space,' *New York Times*, 16 December 1963, p. 33; 'X-Rays and Neutron Stars,' *Sky and Telescope* 27 (1964), p. 215; and 'X-Ray Astronomy,' *Scientific American* 209 (No. 6, 1963), pp. 67–8.

Chapter 5: 'A major area within astronomy'

1. George Gamow, 'The Evolutionary Universe,' in Owen Gingerich, ed., *New Frontiers in Astronomy* (San Francisco: W. H. Freeman, 1975), pp. 316–23; and R. A. Alpher, H. Bethe, and G. Gamow, 'The Origin of Chemical Elements,' *Physical Review* 73 (1948), pp. 803–4.
2. See H. Bondi and T. Gold, 'The Steady State Theory of the Expanding Universe,' *Monthly Notices of the Royal Astronomical Society* 108 (1948), pp. 252–70; and F. Hoyle and J. V. Narlikar, 'Mach's Principle and the Creation of Matter,' *Proceedings of the Royal Society* 273 (1963), pp. 1–11. For a modern treatment of the steady state theory, see Fred Hoyle, *Astronomy and Cosmology* (San Francisco: W. H. Freeman, 1975), pp. 675–81.
3. T. Gold and F. Hoyle, *Paris Symposium on Radio Astronomy*, 1958, cited in F. Hoyle, 'X-Rays From Outside the Solar System,' *Astrophysical Journal* 137 (1963), p. 994.
4. R. J. Gould and G. R. Burbidge, 'X-Rays from the Galactic Center, External Galaxies, and the Intergalactic Medium,' *Astrophysical Journal* 138 (1963), p. 976.
5. Interview with Riccardo Giacconi, 13 February 1979.
6. K. G. Jansky, 'Directional Studies of Atmospherics at High Frequencies,' *Proceedings of the Institute of Radio Engineers* 20 (1932), pp. 1920–32; K. G. Jansky, 'Electrical Disturbances Apparently of Extraterrestrial Origin,' *Proceedings of the Institute of Radio Engineers* 21 (1933), p. 1387; K. G. Jansky, 'Radio Waves from Outside the Solar System,' *Nature* 132 (1933), p. 66; and K. G. Jansky, 'A Note on the Source of Interstellar Interference,' *Proceedings of the Institute of Radio Engineers* 23 (1935), pp. 1158–63.
7. G. Reber, 'Cosmic Static,' *Astrophysical Journal* 91 (1940), pp. 621–4. Also see G. Reber, 'Cosmic Static,' *Astrophysical Journal* 100 (1944), pp. 279–87.
8. A. C. B. Lovell, 'The New Science of Radio Astronomy,' *Nature* 167 (1951), p. 96.
9. J. S. Hey, *The Evolution of Radio Astronomy* (New York: Science History, 1973), p. 55.

10. *Ibid.*, p. 56.

11. J. S. Shklovsky, 'Radio Galaxies,' *Soviet Astronomy – AJ* 4 (1960), pp. 885 – 7; and G. R. Burbidge, 'Galactic Explosions as Sources of Radio Emission,' *Nature* 190 (1961), pp. 1053 – 6.

12. M. Schmidt, '3C273: A Star-Like Object with Large Red-Shift,' *Nature* 197 (1963), p. 1040; and J. B. Oke, 'Absolute Energy Distribution in the Optical Spectrum of 3C273,' *Nature* 197 (1963), pp. 1040 – 1.

13. Hong-Yee Chiu, 'Gravitational Collapse,' in Ivor Robinson, Alfred Schild, and E. L. Schucking, eds., *Quasi-Stellar Sources and Gravitational Collapse* (Chicago: University of Chicago Press, 1965), p. 3. Also see K. N. Douglas, Ivor Robinson, Alfred Schild, E. L. Schucking, J. A. Wheeler, and N. J. Woolf, eds., *Quasars and High Energy Astronomy* (New York: Gordon and Breach, 1969). For popular views of quasars, see Herbert Friedman, *The Amazing Universe* (Washington, D.C.: National Geographic Society, 1975), p. 153; and Maarten Schmidt and Francis Bello, 'The Evolution of Quasars,' in Gingerich, *New Frontiers in Astronomy*, pp. 336 – 46.

14. Hong-Yee Chiu, 'Study of Quasars May Help Solve Mysteries of Universe, *Goddard Institute Bulletin* 8 (No. 9, 1965), p. 8.

15. Bruno Rossi, 'X-Ray and Gamma-Ray Astronomy,' NASA document N69-75301, NASA, Washington, D.C., November 1964, p. 2. Rossi made a similar remark to a *Time* magazine reporter a year earlier. 'X-Rays in the Unknown,' *Time*, 1 November 1963, p. 65.

16. National Academy of Sciences Space Science Board, *Space Research: Directions for the Future*, part 2 (Washington, D.C., National Academy of Sciences, 1966), p. 123.

17. AS&E, Inc., 'An Experimental Program of Extra-Solar X-Ray Astronomy,' proposal to NASA from AS&E, Inc., Cambridge, Massachusetts, 25 September 1963.

18. Prepared testimony of Homer E. Newell, Jr, in US Congress, House, Committee on Science and Astronautics, 1966 *NASA Authorization Hearings*, 89th Congress, 1st session, part 3, p. 526. NASA awarded contract NASW-1506 to AS&E in October 1966 for 'An X-Ray Explorer to Survey Galactic and Extragalactic Sources-Program Definition.'

19. A short history of the HEAO project was provided by John E. Naugle, Associate Administrator for Space Science, in US Congress, House, Committee on Science and Astronautics, 1973 *NASA Authorization Hearings*, 92nd Congress, 2nd session, part 3, p. 52.

20. President's Science Advisory Committee, Joint Space Panels, *The Space Program in the Post-Apollo Period* (Washington, D.C.: Government Printing Office, 1967), p. 29.

21. Robert O. Doyle, ed., *A Long Range Program in Space Astronomy: Position Paper of the Astronomy Missions Board* (Washington, D.C.: NASA, 1969), p. 18.

22. National Academy of Sciences Space Science Board, *Priorities for Space Research*, 1971 – 80 (Washington, D.C.: National Academy of Sciences, 1971), p. 9.

23. Doyle, *A Long Range Program in Space Astronomy*, p. 26.

24. Thomas F. Gieryn of Indiana University kindly provided this unpublished material used in the preparation of his dissertation, 'Patterns in the Selection of Scientific Research Problems: American Astronomers, 1950 – 75,' (Ph.D. dissertation, Columbia University, 1979).

25. Though Gieryn's data base presents some problems for my study of X-ray astronomy, they are not intolerable. First of all, the loss of information on foreign investigators is not critical because the bulk of participants in the field

were American; a study of authors of articles listed in the abstracting journals indicates that most of them worked in the United States (see Chapter 6). Next, Gieryn's data excludes scientists who died before 1975. This problem is also not serious because only one X-ray investigator who began research in the 1960s was no longer alive in the 1970s. Finally, the data include gamma ray researchers (since the field of interest is listed as 'X-Ray and Gamma Ray Astronomy') – people whose work is not generally the subject of this book. Nevertheless, the number of gamma ray investigators included in Gieryn's data cannot be very significant since the activity of all these scientists (American and foreign) never exceeded 29% after 1963. (The percentage fluctuated from 9% in 1972 to 29% in 1965.) One can therefore assume that Gieryn's data on X-ray and gamma ray astronomy represents largely the number of participants working in X-ray astronomy.

Chapter 6: Migrants and money

1. The sample of all American and foreign 'active' experimentalists – those authoring seven or more papers in the field – numbered 98. By choosing seven as the cutoff point, I obtained a manageable sample of names for analysis. Altogether, 832 scientists authored papers in the field between 1962 and 1972. Of these, 414 wrote one paper, 134 wrote two, 69 wrote three, 48 wrote four, 28 wrote five, and 13 wrote six. The distribution for the remaining authors is included in Appendix 1.

2. Interview with Nancy G. Roman, 30 June 1976. Similar views were expressed to me by Arthur D. Code, 9 July 1976; Riccardo Giacconi, 12 July 1976; George W. Clark, 15 July 1976; and George B. Field, 14 July 1980. Of course, a generalization like this concerning the behaviors of traditional astronomers cannot be all encompassing. There were a number of classically trained astronomers who, from the early days of the space program, took advantage of the opportunity to observe the universe through different spectral windows. Included among these were Fred Whipple of the Smithsonian Astrophysical Observatory, Fred Haddock of the University of Michigan, Leo Goldberg of Harvard University, and Arthur Code of the University of Wisconsin. In X-ray astronomy, a good example of a classically trained astronomer who helped design experiments was Hugh M. Johnson of the Lockheed company.

3. P. Morrison, 'On Gamma-Ray Astronomy,' *Nuovo Cimento* 7 (1958), pp. 863–4.

4. Interview with George W. Clark, 20 October 1978. We now know that this range of gamma radiation (below 10 Mev) is the least intense of all.

5. Interview with William L. Kraushaar, 24 October 1978.

6. W. L. Kraushaar and G. W. Clark, 'Search for Primary Cosmic Gamma Rays with the Satellite Explorer XI,' *Physical Review Letters* 8 (1962), pp. 106–9; and W. L. Kraushaar, G. W. Clark, G. Garmire, H. Helmken, P. Higbie, and M. Agogino, 'Explorer XI Experiment on Cosmic Gamma Rays,' *Astrophysical Journal* 141 (1965), pp. 845–63. The low count rate resulted partly from some electronics problems on the satellite. More importantly, however, the gamma ray flux was simply much lower than originally expected.

7. Trevor C. Weekes, *High Energy Astrophysics* (London: Chapman and Hall, 1969), p. 136.

8. Interview with William L. Kraushaar, 24 October 1978. Kraushaar himself continued his collaboration with Clark and others at MIT in designing a gamma ray experiment, flown in 1967 on the third Orbiting Solar Observatory, which revealed gamma rays coming from the galactic plane. G. W. Clark, G.

P. Garmire, and W. L. Kraushaar, 'Observation of High Energy Cosmic Gamma Rays,' *Astrophysical Journal* 153 (1968), pp. L203 – 7.

9. This figure is based on data contained in a letter written by George W. Clark to Nancy Roman of NASA's Office of Space Science and Applications, 28 January 1966. In 1976, these costs had risen to about $30 000 per flight. See National Academy of Sciences Balloon Study Committee, *The Use of Balloons for Physics and Astronomy* (Washington, D.C.: National Academy of Sciences, 1976), p. 10.

10. George W. Clark, 'Balloon Observation of the X-Ray Spectrum of the Crab Nebula Above 15 keV,' *Physical Review Letters* 14 (1965), p. 92.

11. G. Pfotzer, 'History of the Use of Balloons in Scientific Experiments,' *Space Science Reviews* 13 (1972), p. 231. Also see L. E. Peterson and J. R. Winckler, 'Gamma Ray Burst from a Solar Flare,' *Journal of Geophysical Research* 64 (1959), pp. 697 – 707; and J. R. Winckler, L. E. Peterson, R. Hoffman, and R. Arnoldy, 'Auroral X-Rays, Cosmic Rays, and Related Phenomena during the Storm of February 10 – 11, 1958,' *Journal of Geophysical Research* 64 (1959), pp. 597 – 610.

12. See Laurence E. Peterson, Allan S. Jacobson, and R. M. Pelling, 'Spectrum of Crab Nebula X-Rays to 120 keV,' *Physical Review Letters* 16 (1966), pp. 142 – 4; and L. E. Peterson, R. M. Pelling, and J. L. Matteson, 'Techniques in Balloon X-Ray Astronomy,' *Space Science Reviews* 13 (1972), pp. 320 – 36.

13. While the Goddard Space Flight Center group was established within NASA, it operated in a similar fashion to outside university groups. As such, it competed for funding and technical assistance to perform X-ray astronomy experiments. (Interview with Jacob I. Trombka, 7 December 1978.) For a history of the research performed by the Goddard group, see Elihu A. Boldt, Stephen S. Holt, Richard E. Rothschild, and Peter J. Serlemitsos, 'X-Ray Astronomy at Goddard: The First Decade,' Goddard publication X-661-75-122, Goddard Space Flight Center, Greenbelt, Maryland, May 1975.

14. For a summary of these high-energy X-ray observations, see L. E. Peterson, 'Hard Cosmic X-Ray Sources,' in H. Bradt and R. Giacconi, eds., *X-Ray and Gamma Ray Astronomy* (Dordrecht, Holland: D. Reidel, 1973), pp. 51 – 73.

15. The detector is described in D. B. Hicks, L. Ried, Jr, and L. E. Peterson, 'X-Ray Telescope for an Orbiting Solar Observatory,' *IEEE Transactions on Nuclear Science* 12 (1965), pp. 54 – 65.

16. Laurence E. Peterson, Allan S. Jacobson, R. M. Pelling, and Daniel A. Schwartz, 'Observations of Cosmic X-Ray Sources in the 10 – 250 keV Range,' *Canadian Journal of Physics* 46 (1968), pp. S437 – 43.

17. Daniel A. Schwartz, Hugh S. Hudson, and Laurence E. Peterson, 'The Spectrum of Diffuse Cosmic X-Rays: 7.7 – 113 keV,' *Astrophysical Journal* 162 (1970), pp. 431 – 7; and Daniel A. Schwartz, 'The Isotropy of the Diffuse Cosmic X-Rays Determined by OSO-III,' *Astrophysical Journal* 162 (1970), pp. 439 – 44.

18. The study consisted of determining whether articles written by 108 randomly selected American members of the American Physical Society in 1970 were of a theoretical or experimental nature. It was assumed that a paper's content accurately reflected the author's primary research focus on theory or experiment.

19. Interview with George W. Clark, 14 February 1979. For more on the Russian space program and its organization, see William H. Schauer, *The Politics of*

Space: A Comparison of the Soviet and American Programs (New York: Holmes and Meier, 1976).

20. The distribution for the remainder was: nine (15%) in general science publications (such as *Science* and *Nature*); five (8%) in specialized space science journals; and 18 (30%) in popular and miscellaneous publications.

21. Of the remaining papers, the distribution was: 27 (14%) in general science vehicles; two (1%) in specialized space science journals; and 13 (6%) distributed among popular and miscellaneous publications.

22. John M. Logsdon, *The Decision to Go to the Moon: Project Apollo and the National Interest* (Cambridge: MIT Press, 1970), pp. 111–12.

23. President John F. Kennedy to Joint Session of Congress, 25 May 1961, in Robert L. Rosholt, *An Administrative History of NASA, 1958–63* (Washington, D.C.: NASA, 1966), pp. 191–2.

24. Despite its central role in making X-ray astronomy research possible by paying for the required high-altitude vehicles and instruments, the exact level of government funding for the field is practically impossible to document. This results largely from the way NASA maintains financial records. The agency provides data on support for research and development in all its physics and astronomy programs, but it does not list funding by individual fields of science. Even when one identifies the groups of scientists performing research in the specialty, problems arise because NASA awards funds so that the exact use of the money is not always clear. In one form of support, the agency offers grants for basic research and teaching in broad fields. To the University of Wisconsin, for example, NASA provided funds for research in X-ray and gamma ray astronomy. Support for investigations that came out of these wide ranging grants, however, is impossible to account for because the grant agreements did not specify the percentage of funds spent on one or another type of research.

25. The AS&E company also received abundant funding for solar X-ray instrumentation. In fact, it received about $29.5 million for solar X-ray astronomy research – $16.3 million *more* than for nonsolar work. Undoubtedly the experience gained and technology developed for solar studies was helpful for nonsolar efforts. For example, the AS&E scientists used the reflecting X-ray telescope for observing the Sun on a Skylab mission a few years before it was launched on the High Energy Astronomical Observatory.

26. Data obtained from contract records at the NASA Office of Procurement, Washington, D.C.

27. John R. Holtz, NASA Astrophysics Division, Evaluation Report of 'Proposal for Support of the NRL E. O. Hulburt Center for Space Research,' 31 May 1967; and Herbert Friedman, 'E. O. Hulburt Center for Space Research Colloquia for Fall 1967 to Spring 1968,' NRL, Washington, D.C., no date.

28. For a history of the founding of the LRL and LASL, see Herbert F. York, *The Advisors: Oppenheimer, Teller, and the Superbomb* (San Francisco: W. H. Freeman, 1976).

29. Interview with Roderick L. Grader, 9 March 1978.

30. Interview with Frederick D. Seward, 27 March 1978.

31. *Ibid.*

32. Glenn A. Fowler, Randall C. Maydew, and William R. Barton, 'Sandia Laboratories Rocket Program – A Review,' Report SAND 76-0184, Sandia Laboratories, Albuquerque, New Mexico, April 1976, p. 8.

33. Interview with Frederick D. Seward, 27 March 1978.

34. *Ibid.*

35. Interview with Charles D. Swift, 9 March 1978. This large detector is described in R. E. Price, D. J. Groves, R. M. Rodrigues, F. D. Seward, C. D.

Swift, and A. Toor, 'X-Rays from the Magellanic Clouds,' *Astrophysical Journal* 168 (1971), pp. L7 – 9.

36. This view was expressed during interviews with Talbot A. Chubb, 31 May 1978, and William L. Kraushaar, 24 October 1978.

37. G. Chodil, Hans Mark, R. Rodrigues, F. D. Seward, C. D. Swift, Isaac Turiel, W. A. Hiltner, George Wallerstein, and E. J. Mannery, 'Simultaneous Observations of the Optical and X-Ray Spectra of Sco XR-1,' *Astrophysical Journal* 154 (1968), pp. 645 – 53.

38. 'Vela' means 'vigil' in Spanish.

39. For details on the Vela program, see C. M. Beyer, 'History, Organization, and Funding of Project Vela,' in Advanced Research Projects Agency, 'Project Vela: Proceedings of a Symposium,' report TR 60 – 6/U, Department of Defense, Washington, D.C., October 1960, pp. 5 – 26; and US Congress, Joint Committee of Atomic Energy, *Developments in the Field of Detection and Identification of Nuclear Explosions (Project Vela) and Relationship to Test Ban Negotiations,* summary-analysis of hearings held 25 – 7 July 1961 (Washington, D.C.: Government Printing Office, 1962).

40. Sidney Singer, 'Vela High Altitude Program,' *Proceedings of the IEEE* 53 (1965), p. 1935.

41. Interview with Jerry P. Conner, LASL, 13 April 1978; J. P. Glore, 'Vela Satellite System Detector Electronics,' *Proceedings of the IEEE* 53 (1965), p. 1949; R. Stephen White, 'Background Measurements for Vela Hotel,' in US Congress, Joint Committee of Atomic Energy, *Developments in the Field of Detection,* pp. 272 – 3; Bill Richmond, 'Vela Stellites 9 and 10,' *Atom,* July 1969, p. 5; and Bill Richmond, 'Last of the Velas,' *Atom,* May 1970, p. 3. (*Atom* is a publication of LASL.)

42. Jerry P. Conner and Sidney Singer, 'Satellite Observations of Galactic X-Ray Sources,' *Astronomical Journal* 73 (1968), p. S89.

43. J. P. Conner, W. D. Evans, and R. D. Belian, 'The Recent Appearance of a New X-Ray Source in the Southern Sky,' *Astrophysical Journal* 157 (1969), p. L157.

44. *Ibid.,* p. L159.

45. Interview with Frederick D. Seward, 27 March 1978.

46. Interviews with Charles D. Swift, 9 March 1978; and Jerry P. Conner, 13 April 1978.

47. Thomas F. Gieryn and Richard F. Hirsh, 'Marginality and Innovation in Science,' *Social Studies of Science* 13 (1983), pp. 87–106.

Chapter 7: Of mechanisms and a model

1. The mechanism was first outlined in Riccardo Giacconi, Herbert Gursky, Frank R. Paolini, and Bruno B. Rossi, 'Evidence for X – Rays from Sources Outside the Solar System,' *Physical Review Letters* 9 (1962), pp. 442 – 3. In fuller detail, it was presented in G. W. Clark, 'The Relation between Cosmic Gamma Rays and Synchrotron X-Rays,' *Nuovo Cimento* 30 (1963), pp. 727 – 33.

2. J. E. Felten and P. Morrison, 'Recoil Photons from Scattering of Starlight by Relativistic Electrons,' *Physical Review Letters* 10 (1963), pp. 453 – 7. The spectra of inverse Compton and synchrotron radiation are similar because the energy distribution of the generating electrons is the same. See Richard B. Hoover, Roger J. Thomas, and James H. Underwood, 'Advances in Solar and Cosmic X-Ray Astronomy: a Survey of Experimental Techniques and Observational Results,' *Advances in Space Science and Technology* 11 (1972), p. 16.

3. J. W. Overbeck, 'Experimental Test of the Inverse Compton Effect,' (Ph.D.

dissertation, MIT, 1964), cited in Riccardo Giacconi and Herbert Gursky, 'Observation of X-Ray Sources Outside the Solar System,' *Space Science Reviews* 4 (1965), p. 165.

4. For a review of the fate of blackbody radiation as an emission mechanism during the 1960s, see L. Woltjer, 'Emission Mechanisms in X-Ray Sources,' in L. Gratton, ed., *Non-Solar X- and Gamma Ray Astronomy* (Dordrecht, Holland: D. Reidel, 1970), pp. 208–9.

5. T. A. Chubb, H. Friedman, and R. W. Kreplin, 'Measurements Made of High Energy X-Rays Accompanying Three Class 2+ Flares,' *Journal of Geophysical Research* 65 (1960), pp. 1831–2. For a textbook description of the process of thermal bremsstrahlung emitted from a Maxwellian distribution of electrons, see Lyman Spitzer, Jr, *Physics of Fully Ionized Gases*, 2nd ed. (New York: John Wiley and Sons, 1962), pp. 148–9.

6. Bruno B. Rossi, 'Remarks on X-Ray Astronomy,' in *Proceedings of the Solvay Institute 13th Physics Conference*, Brussels, September 1964 (New York: Interscience Publishers, 1965), pp. 124–5.

7. Riccardo Giacconi, 'Introduction,' in Riccardo Giacconi and Herbert Gursky, eds., *X-Ray Astronomy* (Dordrecht, Holland: D. Reidel, 1974), p. 14; and interview with Riccardo Giacconi, 13 February 1979.

8. Interviews with Bruno B. Rossi, 14 February 1979; and George W. Clark, 14 February 1979.

9. David O. Edge and Michael J. Mulkay, *Astronomy Transformed: The Emergence of Radio Astronomy in Britain* (New York: John Wiley and Sons, 1976), pp. 204–5.

10. H. N. Russell, 'Giant and Dwarf Stars,' *Observatory* 463 (August 1913), pp. 324–9; and H. N. Russell, 'Stellar Spectra and Cosmical Evolution,' *Nature* 93 (1914), pp. 227–30, 252–8, and 281–6.

11. R. H. Fowler, 'Dense Matter,' *Monthly Notices of the Royal Astronomical Society* 87 (1926), pp. 114–22; and E. Fermi, 'Zur Quantenlung des idealen einatomigen Gases,' *Zeitschrift fur Physik* 36 (1926), pp. 902–12. For reviews of the theory of dwarfs, see A. S. Eddington, *Stars and Atoms* (New Haven: Yale University Press, 1927), pp. 122–7; and Herbert Gursky and Remo Ruffini 'Introduction,' in Herbert Gursky and Remo Ruffini, eds., *Neutron Stars, Black Holes, and Binary X-Ray Sources* (Dordrecht, Holland: D. Reidel, 1975), pp. 2–3.

12. See, for example, George Gamow and C. L. Critchfield, *Theory of the Atomic Nucleus and Nuclear Energy Sources*, 3rd ed. (Oxford: Oxford University Press, 1949), pp. 301–2.

13. Edwin P. Hubble, *The Realm of the Nebulae* (New Haven: Yale University Press, 1936).

14. Baade and Zwicky gave the supernova its name. W. Baade and F. Zwicky, 'On Supernovae,' *Proceedings of the Academy of Sciences* 20 (1934), pp. 254–9.

15. W. Baade and F. Zwicky, 'Supernovae and Cosmic Rays,' *Physical Review* 45 (1934), p. 138.

16. F. Zwicky, 'On the Theory and Observation of Highly Collapsed Stars,' *Physical Review* 55 (1939), pp. 726–43.

17. A. G. W. Cameron, 'Neutron Star Models,' *Astrophysical Journal* 130 (1959), p. 884; and S. Chandrasekhar, 'The Increasing Role of General Relativity in Astronomy,' *Observatory* 92 (1972), p. 167.

18. S. Chandrasekhar, 'Stellar Configurations with Degenerate Cores,' *Observatory* 57 (1934), pp. 373–7; and S. Chandrasekhar, 'Highly Collapsed Configurations of a Stellar Mass, part II,' *Monthly Notices of the Royal Astronomical Society* 95 (1935), pp. 226–60.

19. Cameron, 'Neutron Star Models,' p. 884.
20. E. Margaret Burbidge, G. R. Burbidge, William A. Fowler, and F. Hoyle, 'Synthesis of the Elements in Stars,' *Reviews of Modern Physics* 29 (1957), pp. 547–650.
21. Cameron, 'Neutron Star Models,' p. 894.
22. J. R. Oppenheimer and G. M. Volkoff, 'On Massive Neutron Cores,' *Physical Review* 55 (1939), pp. 374–81; and J. R. Oppenheimer and H. Snyder, 'On Continued Gravitational Contraction,' *Physical Review* 56 (1939), pp. 455–9. These papers constitute the cornerstones of relativistic astrophysics and gravitational physics. See Gursky, 'Introduction,' p. 6.
23. Cameron, 'Neutron Star Models,' pp. 884–95.
24. Hong-Yee Chiu, 'Selected Topics in Particle Physics,' (Ph.D. dissertation, Cornell University, 1959).
25. Hong-Yee Chiu, *Neutrino Astrophysics* (New York: Gordon and Breach, 1965), p. 93; and Hong-Yee Chiu, 'The Formation of Neutron Stars and Their Surface Properties,' in Ivor Robinson, Alfred Schild, and E. L. Schucking, eds., *Quasi-Stellar Sources and Gravitational Collapse* (Chicago: University of Chicago Press, 1965), p. 424.
26. *Ibid.*
27. Chiu, *Neutrino Astrophysics*, p. 94.
28. The paper was published in 1964 as Hong-Yee Chiu, 'Supernovae, Neutrinos, and Neutron Stars,' *Annals of Physics* 26 (1964), pp. 364–410.
29. While using a detector that could only determine regions of emission to within two degrees, the NRL group was fairly confident that the X-ray source was the extended nebula – having an angular width of about 12 arc minutes – and not another type of celestial body behind it. The confidence was based not only on the fact that the source had been predicted before its discovery, but also on an operational test, which was to become widely used for identifying X-ray emitters in other spectral regions: only unusual objects – certainly different from main sequence stars like the Sun – would emit X-rays in copious quantities. This is essentially the same test used by radio astronomers for identifying optical counterparts. See Edge, *Astronomy Transformed*, p. 103.
30. S. Bowyer, E. T. Byram, T. A. Chubb, and H. Friedman, 'X-Ray Sources in the Galaxy,' *Nature* 201 (1964), p. 1307.
31. S. Bowyer, E. T. Byram, T. A. Chubb, and H. Friedman, 'The X-Ray Sources in Scorpius and the Crab Nebula,' *Naval Research Reviews*, February 1964, p. 4; and Herbert Friedman, 'Cosmic Rays and Gamma Rays,' *Astronautics and Aeronautics* 3 (October 1965), p. 9.
32. The radiation from electrons varies inversely with the 3/2 power of the magnetic field strength multiplied by the square root of the frequency. For the entire expression of the spectrum, see Wallace H. Tucker, *Radiation Processes in Astrophysics* (Cambridge: MIT Press, 1975), p. 111.
33. Friedman, 'Cosmic X-Rays and Gamma Rays,' p. 9.
34. Donald C. Morton, 'Neutron Stars as X-Ray Sources,' *Astrophysical Journal* 140 (1964), p. 469; Donald C. Morton, 'Neutron Stars as X-Ray Sources,' *Nature* 201 (1964), pp. 1308–9; and Donald C. Morton, 'The X-Ray Emission of Neutron Stars,' *Annales d'Astrophysique* 27 (1964), pp. 813–4.
35. S. Bowyer, E. T. Byram, T. A. Chubb, and H. Friedman, 'Galactic X-Ray Astronomy,' *Report of NRL Progress*, February 1964, p. 4.
36. Herbert Friedman, *The Amazing Universe* (Washington, D.C.: National Geographic Society, 1975), p. 100; interviews with C. Stuart Bowyer, 28 July 1976; and Talbot A. Chubb, 31 May 1978.

37. S. Bowyer, E. T. Byram, T. A. Chubb, and H. Friedman, 'Lunar Occultation of X-Ray Emission from the Crab Nebula,' *Science* 146 (1964), p. 912.
38. T. A. Chubb, H. Friedman, and R. W. Kreplin, 'X-Ray Emission Accompanying Solar Flares,' in Naval Research Laboratory, 'Naval Research Laboratory Papers Presented at the Tenth International Astrophysical Symposium, Liege, Belgium; July 11 – 15, 1960,' report 5608, NRL, Washington, D.C., July 1961, pp. 26 – 32.
39. Friedman, *The Amazing Universe*, p. 100.
40. Interview with C. Stuart Bowyer, 28 July 1976.
41. D. S. Stacey, G. A. Stith, R. A. Nidley, and W. B. Pietenpol, 'Rocket Borne Servo Tracks the Sun,' *Electronics* 27 (1954), pp. 149 – 51.
42. Interview with James E. Kupperian, 12 September 1977.
43. The gas supply for the jets came from what remained in the Aerobee rocket's sustainer fuel, oxidizer, and pressurization tanks after burn-out in a manner that made maximum use of the supplies already on board the vehicle. Space General Corporation, 'Aerobee 150 User's Manual,' report SGC 758M-1, Space General Corporation, El Monte, California, July 1966, p. 6-6
44. Interview with William A. Russell, Jr, 1 August 1977. (Russell was the first head of the Attitude Control and Stabilization Branch in the Sounding Rocket Division at the NASA Goddard Space Flight Center.)
45. 'Goddard Aerobee Attitude Control System Record, 1961 – 71,' a record of attitude control systems used on Aerobee rockets managed by the NASA Goddard Space Flight Center, Greenbelt, Maryland.
46. S. Bowyer, E. T. Byram, T. A. Chubb, and H. Friedman, 'The Lunar Occultation of X-Ray Emission from the Crab Nebula,' *Report of NRL Progress*, August 1964, p. 3.
47. S. Bowyer, E. T. Byram, T. A. Chubb, and H. Friedman, 'Rocket Astronomy Studies of the Crab Nebula,' *Transactions of the American Geophysical Union* 45 (1964), p. 562.
48. Interviews with C. Stuart Bowyer, 28 July 1978; and Talbot A. Chubb, 31 May 1978.
49. The counters were eventually patented by Bowyer. Charles S. Bowyer, 'Proportional Counter Tube Having a Plurality of Interconnected Ionization Chambers,' patent 3 396 300, filed 30 December 1965, awarded 6 August 1968.
50. Interview with Talbot A. Chubb, 31 May 1978.
51. Interview with C. Stuart Bowyer, 10 March 1978.
52. Friedman, *The Amazing Universe*, p. 98; and 'Two Neutron Stars Seen,' *Science News Letter*, 9 May 1964, p. 291.
53. George W. Clark, 'Balloon Observation of the X-Ray Spectrum of the Crab Nebula Above 15 keV,' *Physical Review Letters* 14 (1965), pp. 93 – 4.
54. R. Giacconi, H. Gursky, and J. R. Waters, 'Spectral Data from the Cosmic X-Ray Sources in Scorpius and Near the Galactic Center,' *Nature* 207 (1965), p. 573 – 4.
55. G. Chodil, R. C. Jopson, Hans Mark, F. D. Seward, and C. D. Swift, 'X-Ray Spectra from Scorpius (Sco XR-1) and the Sun Observed Above the Atmosphere,' *Physical Review Letters* 15 (1965), p. 607.
56. Hong-Yee Chiu, 'A Review of Pulsars,' *Publication of the Astronomical Society of the Pacific* 82 (1970), p. 363.
57. John N. Bahcall and Richard A. Wolf, 'Neutron Stars: II. Neutrino-Cooling and Observability,' *Physical Review* 140 (1965), p. B1465.

Chapter 8: Research programs

1. M. J. Mulkay, G. N. Gilbert, and S. Woolgar, 'Problem Areas and Research Networks in Science,' *Sociology* 9 (1975), p. 197.
2. Lecture on gamma ray bursts given by Paul A. Gorenstein, Smithsonian Astrophysical Observatory, at the Smithsonian Institution, Washington, D.C., 16 November 1978.
3. David O. Edge and Michael J. Mulkay, *Astronomy Transformed: The Emergence of Radio Astronomy in Britain* (New York: John Wiley and Sons, 1976), pp. 89 and 430.
4. *Ibid.*, pp. 89 – 90.
5. *Ibid.*, pp. 154 – 5.
6. Interview with Talbot A. Chubb, 31 May 1978.
7. S. Bowyer, E. T. Byram, T. A. Chubb, and H. Friedman, 'Cosmic X-Ray Sources,' *Science* 147 (1965), pp. 394 – 8.
8. See, for example, E. T. Byram, T. A. Chubb, and H. Friedman, 'Cosmic X-Ray Sources, Galactic and Extragalactic,' *Science* 152 (1966), pp. 66 – 71; H. Friedman, E. T. Byram, and T. A. Chubb, 'Distribution and Variability of Cosmic X-Ray Sources,' *Science* 156 (1967), pp. 374 – 8; R. Giacconi, H. Gursky, J. R. Waters, G. Clark, and B. Rossi, 'Two Sources of Cosmic X-Rays in Scorpius and Sagittarius,' *Nature* 204 (1964), pp. 981 – 2; H. Gursky, P. Gorenstein, and R. Giacconi, 'The Distribution of Galactic X-Ray Sources From Scorpio to Cygnus,' *Astrophysical Journal* 150 (1967), pp. L75 – 84; Philip C. Fisher, Hugh M. Johnson, Willard C. Jordan, Arthur J. Meyerott, and Loren W. Acton, 'Observations of Cosmic X-Rays,' *Astrophysical Journal* 143 (1966), pp. 207 – 17; and Philip C. Fisher, Willard C. Jordan, Arthur J. Meyerott, Loren W. Acton, and Douglas T. Roethig, 'Resolution of X-Ray Sources at Low Galactic Longitude,' *Nature* 211 (1966), pp. 920 – 3.
9. Bowyer, 'Cosmic X-Ray Sources,' pp. 396 – 7.
10. R. Giacconi, H. Gursky, and J. R. Waters, 'Spectral Data from the Cosmic X-Ray Sources in Scorpius and Near the Galactic Center,' *Nature* 207 (1965), p. 572.
11. Friedman, 'Distribution and Variability of Cosmic X-Ray Sources,' p. 375.
12. *Ibid.*
13. *Ibid.*, p. 377.
14. Fisher, 'Observations of Cosmic X-Rays,' p. 216.
15. H. Friedman and E. T. Byram, 'X-Rays from Sources 3C273 and M87,' *Science* 158 (1967), p. 258.
16. E. T. Byram and H. Friedman, 'X-Ray Quasars,' *Report of NRL Progress*, July 1967, p. 12. The dramatic discovery of extragalactic X-ray sources was contested by the radio astronomer Edward Argyle on the basis that the signals observed from 3C273 were 'within the statistical fluctuations expected from a random distribution.' He argued that it was probable that no sources were observed. The NRL scientists pointed out that the AS&E group had made a similar criticism in an unpublished letter to them. They defended their conclusion, however, by presenting the raw data from the experiment, which they had not done in an earlier paper. Reanalysis of the data in light of the criticism led to the same results. See Edward Argyle, 'X-Rays from Source 3C273,' *Science* 159 (1968), p. 747; and H. Friedman, E. T. Byram, and T. A. Chubb, 'X-Rays from 3C273,' *Science* 159 (1968), pp. 747 – 8.
17. L. L. E. Braes and J. W. Hovenier, 'Supernova Phenomena in OB Associations and Cosmic X-Ray Sources,' *Nature* 209 (1966), pp. 360–1.

18. See, for example, J. S. Shklovsky, 'Remarks on the Nature of X-Ray Sources,' *Soviet Astronomy – AJ* 9 (1965), pp. 224 – 5; P. Morrison and L. Sartori, 'X-Ray Emission from Remnants of Type-I Supernovas,' *Physical Review Letters* 14 (1965), pp. 771 – 5; and S. Hayakawa, M. Matsuoka, and K. Yamashita, 'A Mechanism of X-Ray Emission from Supernova Remnants and Other Active Objects,' *Space Research* 6 (1966), pp. 68 – 79.

19. Friedman, 'Distribution and Variability of Cosmic X-Ray Sources,' p. 378.

20. G. Chodil, Hans Mark, R. Rodrigues, F. Seward, C. D. Swift, W. A. Hiltner, George Wallerstein, and Edward J. Mannery, 'Spectral and Location Measurements of Several Cosmic X-Ray Sources Including a Variable Source in Centaurus,' *Physical Review Letters* 19 (1967), pp. 681 – 3; and R. J. Francey, A. G. Fenton, J. R. Harries, and K. G. McCracken, 'Variability of Centaurus XR-2,' *Nature* 216 (1967), pp. 773 – 4.

21. Edge, *Astronomy Transformed*, pp. 93 – 4.

22. Interviews of former AS&E investigators indicated that the choice of undertaking this program had been well thought out. The published literature and reports written to NASA during the early 1960s substantiate their recollections. In a proposal written in October 1965, for example, Giacconi noted:

    > In past programs large regions of the sky have been explored to search for X-ray sources. A number of sources have been detected; locations, angular sizes, and spectral data have been obtained and published. The experiments which are currently planned and which are proposed here aim at a more detailed study of a small region of the sky with particular emphasis to the precise determination of the characteristics of single sources.

    Riccardo Giacconi, 'A Program of High Angular Resolution Studies of Celestial X-Ray Sources Using Sounding Rockets,' proposal to NASA, AS&E document ASE-1070, AS&E, Inc., Cambridge, Massachusetts, 14 October 1965, p. iv.

23. Minoru Oda, 'High Resolution X-Ray Collimator with Broad Field of View for Astronomical Use,' *Applied Optics* 4 (1965), p. 143.

24. Interview with Paul A. Gorenstein, 14 February 1979.

25. A good description of the instrument is included in Herbert Gursky, 'The Identification of the X-Ray Source in Scorpius,' *Sky and Telescope* 32 (1966), p. 252.

26. M. Oda, G. Clark, G. Garmire, M. Wada, R. Giacconi, H. Gursky, and J. Waters, 'Angular Sizes of the X-Ray Sources in Scorpio and Sagittarius,' *Nature* 205 (1965), p. 554; G. Clark, G. Garmire, M. Oda, M. Wada, R. Giacconi, H. Gursky, and J. R. Waters, 'Positions of Three Cosmic X-Ray Sources in Scorpio and Sagittarius,' *Nature* 207 (1965), p. 584; and H. Gursky, R. Giacconi, P. Gorenstein, and J. R. Waters, 'Analysis of Sounding Rocket Data Regarding Celestial X-Ray Sources,' Final Report of NASA contract NASW-1396, NASA document CR-80112, AS&E document ASE-1408, AS&E, Inc., Cambridge, Massachusetts, September 1966, p. 2.

27. H. Gursky, R. Giacconi, P. Gorenstein, and J. R. Waters, 'A Program of X-Ray Astronomy from Sounding Rockets,' Final Report of NASA contract NASW-1284, NASA document CR-79097, AS&E document ASE-1372, AS&E Inc., Cambridge, Massachusetts, September 1966, p. 38.

28. *Ibid.*, p. 37.

29. Gursky, 'The Identification of the X-Ray Source in Scorpius,' p. 253.

30. Riccardo Giacconi, 'X-Ray Stars,' *Scientific American* 217 (No. 6, 1967), p. 42. A more detailed discussion of the collimator can be found in R. Giacconi,

H. Gursky, and L. P. Van Speybroek, 'Observational Techniques in X-Ray Astronomy,' *Annual Review of Astronomy and Astrophysics* 6 (1968), pp. 388–92; and H. Gursky, R. Giacconi, P. Gorenstein, J. R. Waters, M. Oda, H. Bradt, G. Garmire, and B. V. Sreekantan, 'A Measurement of the Location of the X-Ray Source Sco X-1,' *Astrophysical Journal* 146 (1966), pp. 310–16.

31. A. R. Sandage, P. Osmer, R. Giacconi, P. Gorenstein, H. Gursky, J. R. Waters, H. Bradt, G. Garmire, B. V. Sreekantan, M. Oda, K. Osawa, and J. Jugaku, 'On the Optical Identification of Sco X-1,' *Astrophysical Journal* 146 (1966), p. 317.

32. H. Gursky, R. Giacconi, P. Gorenstein, J. R. Waters, M. Oda, H. Bradt, G. Garmire, and B. V. Sreekantan, 'A Measurement of the Angular Size of the X-Ray Source Sco X-1,' *Astrophysical Journal* 144 (1966), p. 1251.

33. Hugh M. Johnson, 'The Sky Near the Brightest X-Ray Source in Scorpius,' *Astrophysical Journal* 144 (1966), pp. 635–8; and Gursky, 'A Measurement of the Angular Size of the X-Ray Source Sco X-1,' p. 1251.

34. *Ibid.*

35. Interview with George W. Clark, 20 October 1978.

36. Sandage, 'On the Optical Identification of Sco X-1,' p. 317.

37. *Ibid.*, p. 318.

38. Gursky, 'Analysis of Sounding Rocket Data,' p. 2.

39. Gursky, 'The Identification of the X-Ray Source in Scorpius,' p. 255; and Sandage, 'On the Optical Identification of Sco X-1,' pp. 317–20.

40. *Ibid.*, p. 320; and Giacconi, 'X-Ray Stars,' p. 40.

41. See, for example, Kihachiro Ichimura, Goro Ishida, Jun Jugaku, Minoru Oda, Kiyoteru Osawa, and Minoru Shimizu, 'Optical Observations of Sco X-1,' *Publication of the Astronomical Society of Japan* 18 (1966), pp. 469–71; George S. Mumford, 'Three Color Observations of Scorpius X-1, Nova Herculis 1963, Nova GK Persei 1901, and MH Alpha 328-116,' *Astrophysical Journal* 146 (1966), pp. 962–4; Hirosi Yokoi, Toshio Sato, Maski Morimoto, 'An Upper Limit for the Radio Emission from Sco X-1,' *Publication of the Astronomical Society of Japan* 19 (1966), pp. 472–3; B. H. Andrew and C. R. Purton, 'Detection of Radio Emission from Scorpio X-1,' *Nature* 218 (1968), pp. 855–6; and G. Neugebauer, J. B. Oke, E. Beckline, and G. Garmire, 'A Study of Visual and Infrared Observations of Sco X-1,' *Astrophysical Journal* 155 (1969), pp. 1–9.

42. For descriptions of novae as binary systems, see Robert B. Kraft, 'Binary Stars Among Cataclysmic Variables: III. Ten Old Novae,' *Astrophysical Journal* 139 (1964), pp. 457–75; and W. K. Rose, 'Novae,' in Hong-Yee Chiu and Amador Muriel, eds., *Stellar Evolution* (Cambridge: MIT Press, 1972), p. 292.

43. Satio Hayakawa and Masaru Matsuoka, 'Origin of Cosmic X-Rays,' *Progress of Theoretical Physics, Supplement* 30 (1964), pp. 218–9.

44. Burbidge presented a short account of the collective thinking in Geoffrey R. Burbidge, 'Theoretical Ideas Concerning X-Ray Sources,' in *Radio Astronomy and Galactic Systems*, Proceedings of the IAU Symposium 31 (New York: Academic Press, 1967), pp. 465–6. A later account of this meeting is provided in Geoffrey Burbidge, 'Binary Stars as X-Ray Sources,' *Comments on Astrophysics and Space Physics* 3 (1972), pp. 105–6.

45. R. Giacconi, 'X-Ray Astronomy,' *APL Technical Digest* 15 (No. 1, 1976), p. 19.

46. J. S. Shklovsky, 'On the Nature of the Source of X-Ray Emission of Sco XR-1,' *Astrophysical Journal* 148 (1967), pp. L1–4.

47. Shklovsky, 'On the Nature of the Source of X-Ray Emission of Sco XR-1,' pp. L1–4; and J. S. Shklovsky, 'The Nature of the X-Ray Source Sco X-1,'

Soviet Astronomy – AJ 11 (1968), pp. 749–55. For an early criticism of Shklovsky's hypothesis, see Sabatino Sofia, 'Comments on Shklovsky's Model for the X-Ray Source Sco XR-1,' *Astrophysical Journal* 149 (1967), pp. L59–60.

48. H. Friedman, E. T. Byram, and T. A. Chubb, 'Spectrum and Distance of Source Sco XR-1,' *Science* 153 (1966), pp. 1527–8.

49. H. Gursky, R. Giacconi, P. Gorenstein, H. Mako, and J. R. Waters, 'A Program of High Angular Resolution Studies of Celestial X-Ray Sources,' Final Report of NASA contract NASW-1505, NASA document CR-752, p. 2; R. Giacconi, P. Gorenstein, H. Gursky, and J. R. Waters, 'An X-Ray Survey of the Cygnus Region,' *Astrophysical Journal* 148 (1967), pp. L119–27; and R. Giacconi, P. Gorenstein, H. Gursky, P. D. Usher, J. R. Waters, A. Sandage, P. Osmer, and J. V. Peach, 'On the Optical Search for the X-Ray Sources Cyg X-1 and Cyg X-2,' *Astrophysical Journal* 148 (1967), pp. L129–32.

50. *Ibid.*, p. L130.

51. Giacconi, 'An X-Ray Survey of the Cygnus Region,' p. L126.

52. See A. G. W. Cameron and M. Mock, 'Stellar Accretion and X-Ray Emission,' *Nature* 215 (1967), pp. 464–6; and K. H. Prendergast and G. R. Burbidge, 'on the Nature of Some Galactic X-Ray Sources,' *Astrophysical Journal* 151 (1968), pp. L83–8.

53. 'X-Rays from Scorpio,' *Time*, 16 September 1966, p. 78.

54. Robert P. Kraft and Marie-Helene Demoulin, 'On the Remarkable Spectroscopic Complexities of Cyg X-2,' *Astrophysical Journal* 150 (1967), pp. L183–8; and W. A. Hiltner and D. E. Mook, 'Optical Observations of Extrasolar Sources,' *Annual Review of Astronomy and Astrophysics* 8 (1970), pp. 145–55.

Chapter 9: More problems, new lines of research

1. A. Hewish, S. J. Bell, J. D. H. Pilkington, P. F. Scott, and R. A. Collins, 'Observation of a Rapidly Pulsating Radio Source,' *Nature* 217 (1968), pp. 709–13.

2. T. Gold, 'Rotating Neutron Stars as the Origin of Pulsating Radio Sources,' *Nature* 218 (1968), pp. 731–2.

3. *Ibid.*, p. 732.

4. D. W. Richards and J. M. Comella, 'The Period of Pulsar NP 0532,' *Nature* 222 (1969), pp. 551–2.

5. Articles written by Hewish on the discovery of pulsars include Anthony Hewish, 'Pulsars,' *Scientific American* 219 (No. 4, 1968), pp. 25–35; and A. Hewish, 'Pulsars and High Density Physics,' *Science* 188 (1975), pp. 1079–83. Literally hundreds of articles on pulsars were published immediately afer Hewish's discoveries of a pulsar. Already in 1968, the *Astronomischer Jahresbericht* listed 140 articles under the heading 'Pulsars.'

6. David H. Staelin and Edward C. Reifenstein, III, 'Pulsating Radio Sources Near the Crab Nebula,' *Science* 162 (1968), pp. 1481–3; and W. J. Cocke, M. M. Disney, and D. J. Taylor, 'Optical Pulsations in the Crab Nebula Pulsar,' *Nature* 221 (1968), pp. 527–9.

7. Interview with Talbot A. Chubb, 31 May 1978. For a sociological study of how scientists mobilized to search for pulsars, see Steven W. Woolgar, 'The Emergence and Growth of Research Areas in Science with Special Reference to Research on Pulsars,' (Ph. D. dissertation, Cambridge University, 1978).

8. G. Fritz, R. C. Henry, J. F. Meekins, T. A. Chubb, and H. Friedman, 'X-Ray Pulsar in the Crab Nebula,' *Science* 164 (1969), p. 709.

9. H. Bradt, S. Rappaport, W. Mayer, R. E. Nather, B. Warner, M. Macfarland, J. Kristian, 'X-Ray and Optical Observations of the Pulsar NP 0532 in the Crab Nebula,' *Nature* 222 (1969), pp. 728 – 31. The MIT balloon group observed the nebula in the spectral range between 25 and 100 keV in May 1969 and further confirmed the results. F. W. Floyd, I. S. Glass, and H. W. Schnopper, 'Hard X-Rays from the Crab Pulsar,' *Nature* 224 (1969), pp. 50 – 1.

10. See, for example, the report of the Goddard Space Flight Center group. E. A. Holdt, U. D. Desai, S. S. Holt, P. J. Serlemitsos, and R. F. Silverberg, 'Pulsed X-Ray Emission of NP 0532 in March 1968,' *Nature* 223 (1969), pp. 280 – 1. Reanalysis of data from a balloon experiment carried out in 1967 by the Rice University group verified the observations of the X-ray pulsar. G. J. Fishman, F. R. Harnden, Jr, and R. C. Haymes, 'Observation of Pulsed Hard X-Radiation from NP 0532 from 1967 Data,' *Astrophysical Journal* 156 (1969), pp. L107 – 10

11. Fritz, 'X-Ray Pulsar in the Crab Nebula,' p. 711.

12. *Ibid.*

13. Paul Gorenstein and Wallace Tucker, 'Supernova Remnants,' in Owen Gingerich, ed., *New Frontiers in Astronomy* (San Francisco: W. H. Freeman, 1975), p. 261.

14. L. Woltjer, 'X-Rays and Type 1 Supernova Remnants,' *Astrophysical Journal* 140 (1964), pp. 1309 – 12.

15. Because the field strength is proportional to the number of lines of force crossing a unit area, it increases as the inverse square of the shrinking radius. Once having a surface strength of up to 10^4 gauss, the eventual star – 10^5 times smaller than the original body – would possess a magnetic field strength as strong as 10^{14} gauss. See Jeremiah P. Ostriker, 'Nature of Pulsars,' in Gingerich, *New Frontiers in Astronomy*, p. 287.

16. J. S. Shklovsky, 'Pulsar NP 0532 and the Injection of Relativistic Particles into the Crab Nebula,' *Astrophysical Journal* 159 (1970), pp. L77 – L80; and Peter Goldreich and William H. Julian, 'Pulsar Electrodynamics,' *Astrophysical Journal* 157 (1969), pp. 869 – 80.

17. Philip A. Charles and Leonard J. Culhane, 'X-Rays from Supernova Remnants,' *Scientific American* 233 (No. 6, 1975), p. 42.

18. A. Finzi and R. A. Wolf, 'Possible Conversion of Rotational Energy of the Neutron Star in the Crab Nebula into Energy of Relativistic Electrons,' *Astrophysical Journal* 155 (1969), pp. L107 – 8.

19. Geoffrey Burbidge, 'Binary Stars as X-Ray Sources,' *Comments on Astrophysics and Space Physics* 3 (1972), p. 106.

20. Wallace H. Tucker, 'Rotating Neutron Stars, Pulsars, and Cosmic X-Ray Sources,' in L. Gratton, ed., *Non-Solar X- and Gamma Ray Astronomy*, (Dordrecht, Holland: D. Reidel, 1970), p. 202.

21. B. Coppi and A. Ferrari, 'Collective Modes of Plasma Surrounding a Collapsed Star,' *Astrophysical Journal* 161 (1970), pp. L65 – 9; and K. Davidson, F. Pacini, and E. E. Salpeter, 'A Cocoon Pulsar Model for Scorpius X-1,' *Astrophysical Journal* 168 (1971), pp. 45 – 55.

22. H. Friedman, G. Fritz, R. C. Henry, J. P. Hollinger, J. F. Meekins, and D. Sadeh, 'Absence of Pulsar Characteristics in Several X-Ray Sources,' *Nature* 221 (1969), pp. 345 – 7.

23. *Ibid.*, and M. Lampton, C. S. Bowyer, and S. Harrington, 'A Search for Rapid Optical Oscillations in Scorpius XR-1,' *Astrophysical Journal* 162 (1970), pp. 181 – 6.

24. Herbert Friedman, 'Pulsars,' *Astronautics and Aeronautics* 89 (1970), p. 61.

25. Robert J. Gould, 'Origin of Cosmic X-Rays,' *American Journal of Physics* 35 (1967), pp. 381 and 389.
26. R. Giacconi, H. Gursky, J. R. Waters, B. Rossi, G. Clark, G. Garmire, M. Oda, and M. Wada, 'Some Observational Aspects of X-Ray Astronomy,' NASA document CR-69224, AS&E document ASE-1003B, AS&E, Inc., Cambridge, Massachusetts, July 1965, pp. 65–7; and F. Seward, G. Chodil, Hans Mark, C. Swift, and A. Toor, 'Diffuse Cosmic X-Ray Background Between 4 and 40 keV,' *Astrophysical Journal* 150 (1967), pp. 845–50.
27. David O. Edge and Michael J. Mulkay, *Astronomy Transformed: The Emergence of Radio Astronomy in Britain* (New York: John Wiley and Sons, 1976), p. 90.
28. R. J. Gould and G. R. Burbidge, 'X-Rays From the Galactic Center, External Galaxies, and the Intergalactic Medium,' *Astrophysical Journal* 138 (1963), p. 976.
29. James E. Felten and Philip Morrison, 'Omnidirectional Inverse Compton and Synchrotron Radiation from Cosmic Distributions of Fast Electrons and Thermal Photons,' *Astrophysical Journal* 146 (1966), pp. 686–708.
30. A. A. Penzias and R. W. Wilson, 'A Measurement of Excess Antenna Temperature at 4080 Mc/s,' *Astrophysical Journal* 142 (1965), pp. 419–21. Also see the preceding article, R. H. Dicke, P. J. E. Peebles, P. G. Roll, and D. T. Wilkinson, 'Cosmic Blackbody Radiation,' *Astrophysical Journal* 142 (1965), pp. 414–9.
31. Gould, 'Origin of Cosmic X-Rays,' p. 389.
32. R. Giacconi, H. Gursky, and L. P. Van Speybroek, 'Observational Techniques in X-Ray Astronomy,' *Annual Review of Astronomy and Astrophysics* 6 (1968), p. 381.
33. *Ibid.*
34. The latitude effect was predicted in R. J. Gould and D. W. Sciama, 'Cosmic X- and Infrared Rays as Tools for Exploring the Large-Scale Structure of the Universe,' *Astrophysical Journal* 140 (1964), pp. 1634–5.
35. John Gribbon, *Our Changing Universe: The New Astronomy* (New York: E. P. Dutton, 1976), pp. 129–30.
36. George B. Field and Richard C. Henry, 'Free-Free Emission by Intergalactic Hydrogen,' *Astrophysical Journal* 140 (1964), pp. 1002–12.
37. Interview with William L. Kraushaar, 24 October 1978.
38. Herbert Friedman, 'Solar Observations Obtained from Vertical Sounding,' *Reports on Progress in Physics* 25 (1962), pp. 187–8.
39. S. Bowyer, E. T. Byram, T. A. Chubb, and H. Friedman, 'Cosmic X-Ray Sources,' *Science* 147 (1965), pp. 394–8; and H. Friedman, E. T. Byram, and T. A. Chubb, 'Spectrum and Distance of Source Sco XR-1,' *Science* 153 (1966), pp. 1527–8.
40. *Ibid.*
41. Interviews with Dan McCammon, 4 December 1978; and William L. Kraushaar, 24 October 1978.
42. C. S. Bowyer, G. B. Field, and J. Mack, 'Detection of an Anisotropic Soft X-Ray Background Flux,' *Nature* 217 (1968), p. 32.
43. *Ibid.*, p. 34.
44. Interview with C. Stuart Bowyer, 10 March 1978.
45. Bowyer, 'Detection of an Anisotropic Soft X-Ray Background Flux,' p. 34.
46. Interview with Richard C. Henry, 5 December 1978; and letter from Richard C. Henry to Richard F. Hirsh, 12 December 1978. Henry outlined the research program in Richard C. Henry, 'X-Ray Background,' internal NRL report, NRL, Washington, D.C., June 1967.

47. R. C. Henry, G. Fritz, J. F. Meekins, H. Friedman, and E. T. Byram, 'Possible Detection of a Dense Intergalactic Plasma,' *Astrophysical Journal* 153 (1968), pp. L11 – 8. The group also observed soft X-rays from Sco X-1 in this experiment. G. Fritz, J. F. Meekins, R. C. Henry, E. T. Byram, and H. Friedman, 'Soft X-Rays from Scorpius XR-1,' *Astrophysical Journal* 153 (1968), pp. L199 – 202.

48. Henry, 'Possible Detection of a Dense Intergalactic Plasma,' p. L15.

49. Letter from Richard C. Henry to George B. Field, University of California at Berkeley, 17 May 1968.

50. Henry, 'Possible Detection of a Dense Intergalactic Plasma,' p. L14.

51. Herbert Friedman, 'Cosmic X-Rays,' *Nature* 220 (1968), p. 865.

52. 'Mystery of the Missing Mass,' *Time*, 3 May 1968, p. 50.

53. A. N. Bunner, P. L. Coleman, W. L. Kraushaar, D. McCammon, T. M. Palmieri, A. Shilepsky, and M. Ulmer, 'Soft X-Ray Background Flux,' *Nature* 223 (1969), pp. 1222 – 6; and A. N. Bunner, P. L. Coleman, W. L. Kraushaar, and D. McCammon, 'Low Energy Diffuse X-Rays,' *Astrophysical Journal* 167 (1971), pp. L3 – 8.

54. See, for example, A. J. Baxter, B. G. Wilson, and D. W. Green, 'The Diffuse X-Radiation Spectrum Below 4 keV,' *Astrophysical Journal* 155 (1969), pp. L145 – 8; and S. Hayakawa, T. Kato, F. Makino, H. Ogawa, Y. Tanaka, K. Yamashita, M. Matsuoka, S. Miyamoto, M. Oda, and Y. Ogawara, 'A Rocket Observation of Cosmic X-Rays in the Energy Range Between 0.15 and 20 keV,' in Gratton, *Non-Solar X- and Gamma Ray Astronomy*, pp. 121 – 9. The results from Baxter's group at the University of Calgary were discredited, since the group's detectors were contaminated by large fluxes of electrons.

55. Interview with Dan McCammon, 4 December 1978; and Bunner, 'Low Energy Diffuse X-Rays,' pp. L3–8.

56. D. McCammon, A. N. Bunner, P. L. Coleman, and W. L. Kraushaar, 'A Search for Absorption of the Soft X-Ray Diffuse Flux by the Small Magellanic Cloud,' *Astrophysical Journal* 168 (1971), pp. L33 – 7.

57. Dan McCammon, 'A Search for Absroption of the Soft X – Ray Diffuse Flux by the Small Magellanic Cloud,' (Ph.D. dissertation, University of Wisconsin, 1971), p. 3.

58. McCammon, 'A Search for Absorption,' pp. L33 – 7.

59. *Ibid.*, p. L36; and interviews with Dan McCammon, 4 December 1978; William L. Kraushaar, 24 October 1978; and Richard C. Henry, 5 December 1978.

60. McCammon, 'A Search for Absorption,' p. L37; and interview with Dan McCammon, 4 December 1978.

61. *Ibid.*; interview with Richard C. Henry, 5 December 1978; and A. Davidsen, S., Shulman, G. Fritz, J. F. Meekins, R. C. Henry, and H. Friedman, 'Observations of the Soft X-Ray Background,' *Astrophysical Journal* 177 (1972), pp. 629 – 42.

62. *Ibid.*; and George B. Field, 'Intergalactic Matter,' *Annual Review of Astronomy and Astrophysics* 10 (1972), p. 245.

63. L. Woltjer, 'Emission Mechanisms in X-Ray Sources,' in Gratton, *Non-Solar X- and Gamma Ray Astronomy*, p. 208. Also see D. J. Adams, *Cosmic X-Ray Astronomy* (Bristol: Adam Hilger, 1980), pp. 38 – 40.

64. Herbert Friedman, 'X-Rays from Stars and Nebulae,' *Annual Review of Nuclear Science* 17 (1967), p. 321.

65. See, for example, Fritz, 'Soft X-Rays from Scorpius XR-1,' pp. L199–202; R. W. Hill, R. J. Grader, and F. D. Seward, 'The Soft X-Ray Spectrum of Sco XR-1,' *Astrophysical Journal* 154 (1968), pp. 655–60; P. Gorenstein, R.

Giacconi, and H. Gursky, 'The Spectra of Several X-Ray Sources in Cygnus and Scorpio,' *Astrophysical Journal* 150 (1967), pp. L85–94; P. Gorenstein, H. Gursky, and G. Garmire, 'The Analysis of X-Ray Spectra,' *Astrophysical Journal* 153 (1968), pp. 885–97; and S. S. Holt, E. A. Boldt, and P. J. Serlemitsos, 'Search for Line Structure in the X-Ray Spectrum of Sco X-1,' *Astrophysical Journal* 158 (1969), pp. L155–8.

66. Friedman, 'X-Rays from Stars and Nebulae,' p. 321.

67. J. R. P. Angel, R. Novick, P. Vanden Bout, and R. Wolff, 'Search for Polarization in Sco X-1,' *Physical Review Letters* 22 (1969), pp. 861–5.

68. H. W. Schnopper, H. V. Bradt, S. Rappaport, E. Boughan, B. Burnett, R. Doxsey, W. Mayer, and S. Watt, 'Precise Location of Sagittarius X-Ray Sources with a Rocket-Borne Rotating Modulation Collimator,' *Astrophysical Journal* 161 (1970), pp. L161–7. Because of the high density of stars in the Sagittarius constellation, scientists found no clear optical counterparts. W. Kunkel, P. Osmer, M. Smith, A. Hoag, D. Schroeder, W. A. Hiltner, H. Bradt, S. Rappaport, and H. W. Schnopper, 'An Optical Search for the X-Ray sources GX3+1, GX5–1, and GX17+2,' *Astrophysical Journal* 191 (1970), pp. L169–72.

69. Giacconi, 'Observational Techniques in X-Ray Astronomy,' p. 391.

70. H. Friedman, 'Cosmic X-Ray Observations,' *Proceedings of the Royal Society* 313A (1969), p. 301.

71. Letter from Homer E. Newell, Jr, NASA Associate Administrator for Space Science and Applications, to Philip C. Fisher, 15 February 1965. For details of the X-ray detector, see P. C. Fisher and A. J. Meyerott, 'High Sensitivity Detectors for Few keV X-Rays,' *IEEE Transactions on Nuclear Science* NS-13 (1966), p. 585.

72. Interview with Loren W. Acton, 10 March 1978.

73. 'OAO Silent,' *Flight International* 89 (1966), p. 677; and memorandum to files of Alan H. Sures, NASA OAO Program Manager, on OAO-I failure, 5 May 1966.

Chapter 10: Uhuru

1. G. Burbidge, 'Binary Stars as X-Ray Sources,' *Comments on Astrophysics and Space Physics* 3 (1972), p. 105.

2. The spacecraft and its mission are described in R. Giacconi, E. Kellogg, P. Gorenstein, H. Gursky, and H. Tananbaum, 'An X-Ray Scan of the Galactic Plane from Uhuru,' *Astrophysical Journal* 165 (1971), pp. L27–35.

3. A history of the Uhuru project is included in Riccardo Giacconi, 'Introduction,' in Riccardo Giacconi and Herbert Gursky, eds., *X-Ray Astronomy* (Dordrecht, Holland: D. Reidel, 1974), p. 20.

4. 'Italy to Launch US Satelite,' NASA News Release 70-203, NASA, Washington, D.C., 2 December 1970, p. 5.

5. Giacconi, 'An X-Ray Scan of the Galactic Plane from Uhuru,' p. L28; and Herbert Gursky and Daniel Schwartz, 'Observational Techniques,' in Giacconi, *X-Ray Astronomy*, pp. 93–4.

6. The principle of pulse shape discrimination was described in 1964 in E. Mathieson and P. W. Sanford, 'Pulse Shape Discrimination in Proportional Counters,' in *Nuclear Electronics* (Paris: ENEA, 1964), pp. 65–9. An abstract of the article appears in *Nuclear Science Abstracts* 18 (1964), p. 3238. The anticoincidence technique was then used in a solar X-ray experiment carried out on a satellite. See J. L. Culhane, J. Herring, P. W. Sanford, G. O'Shea, and R. D. Phillips, 'Advances in the Design and Performance of X-Ray Proportional Counters,' *Journal of Scientific Instruments* 43 (1966),

pp. 908–12. Gorenstein described his use of the technique in nonsolar experiments in Paul Gorenstein and Stanley Mickiewicz, 'Reduction of Cosmic Background in an X-Ray Proportional Counter through Rise-Time Discrimination,' *Review of Scientific Instruments* 39 (1968), pp. 816–20; and P. Gorenstein, E. M. Kellogg, and H. Gursky, 'The Spectrum of Diffuse Cosmic X-Rays, 1–13 keV,' *Astrophysical Journal* 156 (1969), pp. 315–24.

7. Gursky, 'Observational Techniques,' p. 95; and A. C. Fabian, 'Uhuru – The First X-Ray Astronomy Satellite,' *Journal of the British Interplanetary Society* 28 (1975), p. 343.

8. W. Forman, C. Jones, L. Cominsky, P. Julien, S. Murray, G. Peters, H. Tananbaum, and R. Giacconi, 'The Fourth Uhuru Catalog of X-Ray Sources,' *Astrophysical Journal Supplement Series* 38 (1978), pp. 357–412.

9. *Ibid.*, pp. 360–402.

10. G. Chodil, Hans Mark, R. Rodrigues, F. Seward, C. D. Swift, W. A. Hiltner, George Wallerstein, and Edward J. Mannery, 'Spectral and Location Measurements of Several Cosmic X-Ray Sources Including a Variable Source in Centaurus,' *Physical Review Letters* 19 (1967), pp. 681–3.

11. R. Giacconi, H. Gursky, E. Kellogg, E. Schreier, and H. Tananbaum, 'Discovery of Periodic X-Ray Pulsations in Centaurus X-3 from Uhuru,' *Astrophysical Journal* 167 (1971), pp. L67–73.

12. *Ibid.*, p. L72. For reasons why the source was required to be a small object, but not a white dwarf, see Jeremiah P. Ostriker, 'Recent Developments in the Theory of Degenerate Dwarfs,' *Annual Review of Astronomy and Astrophysics* 9 (1971), pp. 353–6.

13. E. Schreier, R. Levinson, H. Gursky, E. Kellogg, H. Tananbaum, and R. Giacconi, 'Evidence for the Binary Nature of Centaurus X-3 from Uhuru X-Ray Observations,' *Astrophysical Journal* 172 (1972), p. L79.

14. *Ibid.*, p. L84.

15. W. Krzeminski, 'The Identification and UBV Photometry of the Visible Component of the Centaurus X-3 Binary System,' *Astrophysical Journal* 192 (1974), pp. L135–8.

16. H. Tananbaum, H. Gursky, E. M. Kellogg, R. Levinson, E. Schreier, and R. Giacconi, 'Discovery of a Periodic Binary X-Ray Source in Hercules from Uhuru,' *Astrophysical Journal* 174 (1972), pp. L143–9.

17. Arthur Davidson, J. Patrick Henry, John Middleditch, and Harding E. Smith, 'Identification of the X-Ray Pulsar in Hercules: A New Optical Pulsar,' *Astrophysical Journal* 177 (1972), pp. L97–102.

18. M. Oda, P. Gorenstein, H. Gursky, E. Kellogg, E. Schreier, H. Tananbaum, and R. Giacconi, 'X-Ray Pulsations from Cygnus X-1 Observed from Uhuru,' *Astrophysical Journal* 166 (1971), pp. L1–7; and E. Schreier, H. Gursky, E. Kellogg, H. Tananbaum, and R. Giacconi, 'Further Observations of the Pulsating X-Ray Source Cygnus X-1 from Uhuru,' *Astrophysical Journal* 170 (1971), pp. L21–7.

19. S. Rappaport, R. Doxsey, and W. Zaumen, 'A Search for X-Ray Pulsations from Cygnus X-1,' *Astrophysical Journal* 168 (1971), pp. L43–7.

20. Paul Murdin and B. Louise Webster, 'Optical Identification of Cygnus X-1,' *Nature* 233 (1971), p. 110.

21. B. Louise Webster and Paul Murdin, 'Cygnus X-1 – A Spectroscopic Binary with a Heavy Companion,' *Nature* 235 (1972), pp. 37–8. Also see H. Tananbaum, H. Gursky, E. Kellogg, R. Giacconi, and C. Jones, 'Observation of a Correlated X-Ray – Radio Transition in Cygnus X-1,' *Astrophysical Journal* 177 (1972), pp. L5–10.

22. Riccardo Giacconi, 'Progress in X-Ray Astronomy,' 34th Richtmyer

Memorial Lecture, Center for Astrophysics Preprint no. 304, Smithsonian
Astrophysical Observatory, Cambridge, Massachusetts, 30 January 1975, p. 17.

23. *Ibid.*; Remo Ruffini, 'The Physics of Gravitationally Collapsed Objects,' in
Herbert Gursky and Remo Ruffini, eds., *Neutron Stars, Black Holes and Binary
X-Ray Sources* (Dordrecht, Holland: D. Reidel, 1975), p. 87; and Herbert
Gursky and Ethan Schreier, 'The Galactic X-Ray Sources,' in *ibid.*, p. 191.

24. The papers that established the foundation of the well-accepted binary
accretion model consist of J. E. Pringle and M. J. Rees, 'Accretion Disc
Models for Compact X-Ray Sources,' *Astronomy and Astrophysics* 21 (1972),
pp. 1–9; Kris Davidson and Jeremiah P. Ostriker, 'Neutron Star Accretion in
a Stellar Wind: Model for a Pulsed X-Ray Source,' *Astrophysical Journal* 179
(1973), pp. 585–98; and N. I. Shakura and R. A. Sunyaev, 'Black Holes in
Binary System. Observational Appearance,' *Astronomy and Astrophysics* 24
(1973), pp. 337–55. Even though two of these papers were published in 1973,
they existed as preprints in 1972. The papers were cited in other published
literature with the following frequencies:

Number of citations in each year

	1973	1974	1975	1976	1977
Pringle	21	21	43	42	45
Davidson	19	26	29	39	25
Shakura	—	35	35	46	42

High citation frequencies over a period of a few years usually indicate that a
paper includes highly regarded fundamental contributions. Citation data
courtesy of Henry Small, Institute for Scientific Information, Philadelphia,
Pennsylvania. For more information on the use of citation data, see Eugene
Garfield, Morton V. Malin, and Henry Small, 'Citation Data as Science
Indicators,' in Yehuda Elkana, Joshua Lederberg, Robert K. Merton, Arnold
Thackray, and Harriet Zuckerman, eds., *Toward a Metric of Science: The
Advent of Science Indicators* (New York: John Wiley and Sons, 1978), pp.
179–207.

25. Davidson, 'Neutron Star Accretion in a Stellar Wind,' pp. 595–6.

26. Pringle, 'Accretion Disc Models for Compact X-Ray Sources,' p. 7.

27. Kip S. Thorne, 'Gravitational Collapse,' *Scientific American* 217 (No. 5, 1967),
pp. 90–1; and Remo Ruffini and John A. Wheeler, 'Introducing the Black
Hole,' *Physics Today* 24 (January 1971), pp. 30–41.

28. N. I. Shakura, 'Black Holes in Binary Systems,' pp. 347–50.

29. *Ibid.*, p. 340.

30. Robert W. Leach and Remo Ruffini, 'On the Masses of X-Ray Sources,'
Astrophysical Journal 180 (1973), pp. L15–18.

31. Ruffini, 'The Physics of Gravitationally Collapsed Objects,' p. 84.

32. E. P. J. van den Heuvel and J. Heise, 'Centaurus X-3, Possible Reactivation
of an Old Neutron Star by Mass Exchange in a Close Binary,' *Nature Physical
Science* 239 (1972), pp. 67–9; E. P. J. van den Heuvel, 'Wolf–Rayet Systems
and the Origin of Massive X-Ray Binaries,' *Nature Physical Science* 242
(1973), pp. 71–2; and E. P. J. van den Heuvel and C. De Loore, 'The Nature
of X-Ray Binaries: III. Evolution of Massive Close Binaries with One
Collapsed Component – with a Possible Application to Cygnus X-3,'
Astronomy and Astrophysics 25 (1973), pp. 287–95. A popularized and slightly
revised version of this theory is presented in Herbert Gursky and Edward P. J.

van den Heuvel, 'X-Ray Emitting Double Stars,' in Owen Gingerich, ed., *New Frontiers in Astronomy* (San Francisco: W. H. Freeman, 1975), pp. 268–79.

33. *Ibid.*, p. 278. Also see W. Tucker and G. Blumenthal, 'Compact X-Ray Sources,' *Annual Review of Astronomy and Astrophysics* 12 (1974), pp. 36–8.

34. The association of massive blue supergiants with X-ray sources is common. By 1975 eight binary X-ray systems had been detected. In five of these, the companions were blue supergiants. Gursky, 'X-Ray Emitting Double Stars,' p. 271.

35. *Ibid.*, p. 278.

36. P. C. Joss, 'Helium Burning Flashes on an Accreting Neutron Star: A Model for X-Ray Burst Sources,' *Astrophysical Journal* 225 (1978), pp. L123–7.

37. See, for example, Herbert Friedman, 'X-Ray Astronomy,' *New Scientist* 28 (1965), p. 906; and K. A. Pounds, 'X-Ray Astronomy,' *Science Journal* 6 (No. 4, 1970), p. 66.

38. George W. Clark, 'X-Ray Astronomy from Uhuru to HEAO-1,' *Journal of the Washington Academy of Sciences* 71 (March 1981), p. 17.

39. For an idea of the extent of observations carried out with the observatory, see Riccardo Giacconi, ed., *X-Ray Astronomy with the Einstein Satellite* (Dordrecht, Holland: D. Reidel, 1981).

40. Interview with Riccardo Giacconi, 13 February 1979.

Chapter 11: Success and frustration

1. M. Ryle, 'Radio Astronomy,' *Reports on Progress in Physics* 13 (1950), p. 186.

2. Letter from Dr. Thomas O. Payne, NASA Administrator, to Senator Clinton P. Anderson, Chairman, Senate Committee on Aeronautical Sciences, 21 November 1969. The figure is based on costs up to 31 July 1969 and includes $2 billion worth of hardware that had been developed by 1969, but which was used after that date. By the end of the Apollo program in December 1972, the cost had risen to $23.5 billion. Frank W. Anderson, Jr, *Orders of Magnitude: A History of the NACA and NASA, 1915–1980*, 2nd edn., (Washington, D.C.: NASA, 1981), p. 81.

3. NASA contract NASW-1284, awarded to AS&E, Inc., August 1965. Data provided by NASA Office of Procurement, Washington, D.C. The experiment was also made possible with funds provided by NASA and channeled through the MIT group. The $362 000 cost for the experiment, however, does not include any of the MIT funds.

4. NASA, 'Mission Operation Report,' for SAS-1, Explorer 42, NASA, Washington, D.C., 1970.

5. The committee members also needed to discuss requests for construction of facilities and program management. See 'Chronological History, Fiscal Year 1970, Budget Submission,' NASA Office of Administration, Budget Operations Division, final report, 12 October 1970. The entire House Committee on Science and Astronautics only met twice, but its Subcommittees on Manned Space Flight, Space Science and Applications, and Advanced Research and Technology met eight, eight, and five times respectively. See US Congress, House, Committee on Science and Astronautics, 1970 *NASA Authorization Hearings*, 91st Congress, 1st session, parts 1, 2, 3, and 4.

6. In the 1970 Authorization hearings, NASA reported allocating $2 976 000 for all astronomy activities in fiscal year 1969. Obviously not all of this money went to support X-ray astronomy. *Ibid.*, part 2, p. 665.

7. Testimony from John E. Naugle, Associate Administrator of the NASA Office of Space Science and Applications, 9 March 1971, in US Congress, House,

Committee on Science and Astronautics, 1972 *NASA Authorization Hearings*, 92nd Congress, 1st session, p. 176; and National Academy of Sciences, *Priority for Space Research*, 1971–80, pp. 19 and 31, contained in *ibid*, pp. 227 and 239.

8. See Daniel J. Kevles, *The Physicists: the History of a Scientific Community in Modern America* (New York: Alfred K. Knopf, 1978), pp. 393–409.

9. See a discussion on the history of the fiscal year budget in US Congress, Senate, Committee on Aeronautical Sciences, *NASA Authorization for Fiscal Year* 1974, 93rd Congress, 1st session, report no. 93-179.

10. Plans for the future were discussed at the Uhuru Memorial Symposium on 13 December 1980. The proceedings were published as the entire issue of the *Journal of the Washington Academy of Sciences* 71 (No. 2, 1981).

Index